WHITE STAR PUBLISHERS

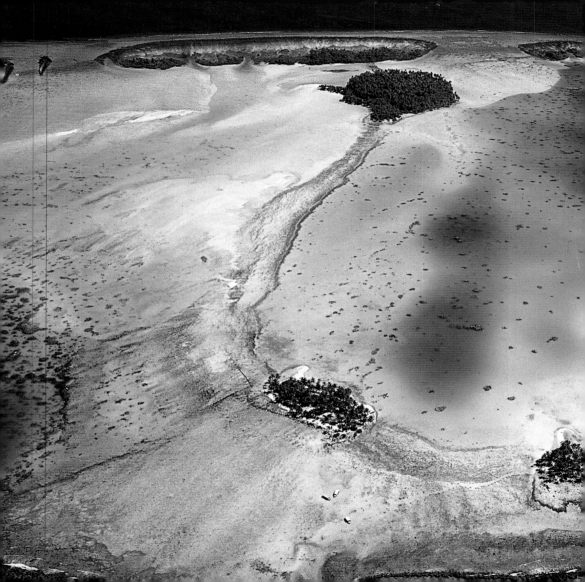

TEXTS BY

ALBERTO BERTOLAZZI

**project manager
and editorial director**
VALERIA MANFERTO DE FABIANIS

graphic design
CLARA ZANOTTI

graphic layout
MARIA CUCCHI

translation
text: RICHARD PIERCE
captions: TIMOTHY STROUD

© 2004 WHITE STAR S.R.L.
VIA CANDIDO SASSONE, 22-24
13100 VERCELLI - ITALY
WWW.WHITESTAR.IT

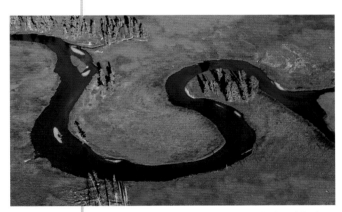

● Wyoming (USA) - Yellowstone National Park.

ISBN 88-544-0001-7

REPRINTS:
1 2 3 4 5 6 08 07 06 05 04

Printed in Singapore
Color separation: Fotomec, Turin, Italy and Chiaroscuro, Turin, Italy

CONTENTS

THE EARTH

PREFACE .... page 18

INTRODUCTION .... page 20

THE PILLARS OF THE SKY .... page 32

THE BLUE DEPTHS .... page 128

SILENT PLANET .... page 226

ROOTS OF THE WORLD .... page 290

A SENSE OF WATER .... page 356

THE MIRRORS OF THE SKY .... page 416

THE GODS OF FIRE .... page 470

THE ICE KINGDOM .... page 498

THE WORLD OF STONE .... page 532

THE WORLD IN BETWEEN .... page 590

INVISIBLE BOUNDARIES .... page 616

ROLLING HILLS .... page 636

BETWEEN LAND AND WATER .... page 660

THE RAINBOW PLANET .... page 696

BIOGRAPHY, INDEX, CREDITS .... page 728

1 ● Red Sea (Egypt) - Ras Gharib.

2-3 ● Rangiroa (French Polynesia) - Blue Lagoon.

4-5 ● Trentino Alto Adige (Italy) - Brenta Dolomites.

6-7 ● Congo - Mayombe Forest.

8-9 ● Utah-Arizona (USA) - Monument Valley.

13 ● Alaska (USA) - Copper River Valley.

14-15 ● Tanzania - Lake Natron.

16-17 ● Kenya - Lake Nakuru.

Preface

THERE IS THE LAND CREATED BY MAN, A LITTLE NEGLECTED BUT LOVED, AND THERE IS THE NATURAL WORLD OF PHYSICAL FORCES AND ENVIRONMENTAL BALANCES. THE FIRST HAS BEEN MOLDED TO BECOME THE UNIQUE EXPRESSION OF THE MOST EVOLVED FORM OF LIFE, THE SECOND IS A SUBLIME MECHANISM, A SYMPHONY IN PERFECT HARMONY, A SCALE IN PERFECT EQUILIBRIUM. THIS BOOK DESCRIBES THE WORLD IN WHICH MAN IS NOT THE CENTRAL CHARACTER BUT SIMPLY AN OBSERVER: NO ROADS, CITIES OR MONUMENTS, BUT WOODS, MOUNTAINS, SEAS, DESERTS AND MUCH MORE. ENCHANTING PORTRAITS OF A WORLD WHERE WE ARE THE GUESTS, NOT THE HOSTS.

Nepal - Ama Dablam.

Introduction

THE ANCIENT AND CONTEMPORARY WORLDS ARE SEPARATED BY A CULTURAL AND TEMPORAL GAP THAT IS NOT EASY TO COMPREHEND. IN THE CLASSICAL AGE THE EARTH WAS A MYSTERIOUS AND UNTOUCHABLE DIVINITY, WHILE NOW, AT THE DAWN OR THE THIRD MILLENNIUM A.D., IT IS BEING STUDIED AND EXPLOITED EVER MORE INTENSIVELY. DURING THE COURSE OF HISTORY, MAN HAS LEARNED TO FORCE OPEN THE "SECRET VAULTS" OF THE PLANET; CONSEQUENTLY, NEVER BEFORE HAS A SPECIES HAD SUCH POWER AND ACHIEVED SUCH A LEVEL OF KNOWLEDGE. BETWEEN THE MIDDLE AGES AND THE MODERN ERA MAN DISCOVERED THE WORLD – SOMETIMES BY INVENTING IT – FIRST THANKS TO FANTASTIC

● Wyoming (USA) - Yellowstone National Park.

Introduction

TALES OF JOURNEYS, THEN THROUGH DIRECT EXPERI-
ENCE AND READING, AND LASTLY BY MEANS OF RAPID
TRANSPORT AND COMMUNICATION. TRAVELERS HAVE
SEEN EVERYTHING THERE IS TO BE SEEN ON THE EARTH,
HAVE TRAVERSED ALL ITS OCEANS, CLIMBED ALL ITS
MOUNTAINS, AND VISITED ALL ITS VILLAGES, DISCOVER-
ING THE STRANGEST AND MOST FASCINATING THINGS,
WHICH THEY HAVE DESCRIBED IN TRAVEL JOURNALS
AND NOVELS. THIS IMMENSE STORE OF ACCOUNTS,
ENHANCED OVER THE CENTURIES BY ALL KINDS OF ILLUS-
TRATIONS, HAS GIVEN RISE TO A DIFFERENT CONSCIOUS-
NESS. THE EARTH, OUR WORLD, HAS ACQUIRED DEPTH,
COLOR AND TASTE, WHILE AT THE SAME TIME SHEDDING
ITS EVERY MYSTERY. THE TREASURES OF NATURE, THE

Introduction

WORKS OF MAN, THE VERY SENSE OF EXISTENCE ON THIS PLANET HAVE ALL BECOME A UNIVERSAL HERITAGE. THIS REVOLUTION, WHICH HAS INFLUENCED HUMAN PERCEPTION AS MUCH AS THE COPERNICAN REVOLUTION, HAS GENERATED THE MODERN URGE TO "TRAVEL WITH OUR EYES WIDE OPEN." WE ARE NO LONGER THE SLAVES, BUT THE MASTERS OF TRAVEL, AND GO AROUND THE WORLD WITH THE SOLE AIM OF INCREASING OUR STORE OF KNOWLEDGE: THE MOST SURPRISING EXPRESSIONS OF LIFE AND NATURE AWAIT US IN PLACES THAT WERE ONCE REMOTE BUT ARE NOW PERHAPS ONLY EXOTIC.... KNOWLEDGE COMES THROUGH OBSERVATION, THAT IS TO SAY, BY STORING IN OUR MEMORY AND THEN REFLECTING AND COMPARING. BY TRAVEL-

Introduction

ING AND OBSERVING, MAN EXAMINES HIMSELF, HE RE-ELABORATES HIS VISION OF THE WORLD BY TAKING IN NEW ASPECTS, EVER MORE REFINED NUANCES. OBSERVATION ESTABLISHES A RELATION WITH THE OBJECT OBSERVED, AND IF THAT OBJECT IS NATURE, THAT MIRACLE OF MIRA-CLES IN THE WORLD, ONE CANNOT BUT BE SPELLBOUND BY IT. THERE ARE IMAGES, ESPECIALLY THE ONES IN THIS BOOK, THAT HELP US TO UNDERSTAND, KNOW AND HENCE LOVE THE ENVIRONMENT IN WHICH WE LIVE: THE MOUN-TAINS, DESERTS, VOLCANOES AND SEAS THAT RELATE TO US THE HISTORY AND FUTURE OF A CELESTIAL BODY AND OF THE FORTUNATE CREATURES THAT LIVE IN IT.

25 ● Fezzan (Libya) - Sahara Desert.

26-27 ● Paranà (Brazil) - Iguazú Falls.

28-29 ● Weddell Sea (Antarctica) - Iceberg in a storm.

30-31 ● Sichuan (China) - Huanglong Nature Park.

The PILLARS of the SKY

Karakorum (Pakistan) - K2.

INTRODUCTION The Pillars of the Sky

A MOUNTAIN CANNOT BE EXPLAINED. IT IS NOT A MERE "IRREGULARITY" OF THE EARTH, BUT A DRAMATIC SYMBOL OF AN ASPIRATION THAT ONLY THE TRUE ENTHUSIAST HAS BEEN ABLE TO DESCRIBE. THE SURVIVORS OF THE MOST DIFFICULT CLIMBS, THOSE WHO WENT BEYOND THE CONFINES OF DEATH, SPEAK OF THEIR BEWILDERMENT BEFORE THE DIVINITY OF STONE AND GLACIERS, HUGE STRETCHES LIKE WHITE DESERTS POPULATED BY DEMONS OF THE IMAGINATION. THESE GIANTS OF THE EARTH HAVE ERECTED FORMIDABLE BARRIERS IN FRONT OF THE FEAR OR ARROGANCE OF THE MANY PERSONS WHO HAVE TRIED TO SCALE THEM, DEMANDING A HUGE TRIBUTE OF BLOOD. ACCORDING TO THE SHERPA, AT TIMES THE MOUNTAINS HAVE EVEN SPOKEN. THIS

INTRODUCTION The Pillars of the Sky

POPULATION IN NEPAL LIVES IN SYMBIOSIS WITH MOUNTAINS, TREATING THEM LIKE DIAMONDS IN THE ROUGH, LIKE SEEDS FROM WHICH ANY FLOWER MIGHT SPRING FORTH, LIKE AN IRASCIBLE AND BEAUTIFUL GOD THAT IS TO BE DEVOUTLY WORSHIPPED. THERE ARE RANGES WHOSE IDENTITY LIES IN THEIR ALTITUDE. THE HIMALAYAS, THE ROOF OF THE EARTH, ARE THE QUINTESSENCE OF THE HIGH MOUNTAIN. THE ANDES, BUFFETED BY THE MOST UNPREDICTABLE BAD WEATHER, SEEM TO MIRROR THE HARSHNESS AND GRANDIOSITY OF THE SOUTH AMERICAN LANDSCAPE. THE ANTARCTIC MOUNTAINS ARE FRAGMENTS OF OTHER WORLDS TOSSED ONTO OUR PLANET FROM ON HIGH; THEY HOST STORMS THAT MAY LAST FOR MONTHS, THEY ARE BATTERED BY CYCLOPEAN

The Pillars of the Sky
Introduction

WINDS, THEY TOWER OVER DESOLATE AREAS THAT LOOK LIKE EXTRATERRESTRIAL DESERTS. THE MOUNTAINS OF NORTHERN CALIFORNIA ARE MONUMENTS CARVED OUT OF ANCIENT ROCK, WITH THE MOST INCONCEIVABLE COLORS – RED, YELLOW, ORANGE, OCHER – STANDING SHEER OVER EXPANSES OF WASTELAND.

MOUNTAINS ARE TRUE SPECTACLE. THOSE WHO CLIMB THEM AFTER HAVING OBSERVED THEM FROM AFAR, EXPERIENCE AT EVERY STEP THAT SUBTLE TINGLE THAT EVERY PERSON FEELS WHEN ENCOUNTERING UNEX-PLORED REGIONS. EVERY CORNER HAS ITS OWN THRILLING SECRET, EVERY PATH CONTAINS THE JOY OF DISCOVERING A BIT OF STILL UNKNOWN NATURE.

37 ● Trentino Alto Adige (Italy) - Brenta Group, Dolomites.

38-39 ● Trentino Alto Adige (Italy) - Catinaccio Group, Dolomites.

Alto Adige (Italy) - Sass
Maor and Cima
Madonna, Dolomites.

42 ● Trentino Alto Adige (Italy) - Cima Tosa, Brenta Group, Dolomites.

43 ● Trentino Alto Adige (Italy) - Crozzon di Brenta, Brenta Group, Dolomites.

44-45 • Piedmont (Italy) - Monte Rosa.

46-47 • Valle d'Aosta (Italy) - The Matterhorn.

48-49 • Valle d'Aosta (Italy) - The Giant's Tooth, Mont Blanc massif.

50-51 • Haut Savoie (France) - Mont Blanc massif.

52-53 • Haut Savoie (France) - Aiguille Noire, Mont Blanc massif.

54-55 • Haut Savoie (France) - Mont Blanc.

56-57 ● Bernese Oberland (Switzerland) - The Eiger, Jungfrau and Mönch.

58-59 ● Bernese Oberland (Switzerland) - south faces of the Eiger and the Mönch.

60-61 ● Bavaria (Germany) - Watzmann (right) and Little Watzmann, Bavarian Alps.

62-63 ● Pyrenees (Spain) - Monte Perdido, Ordesa National Park.

64-65 ● Sichuan (China) - Siguniang.

66-67 ● Tibet (China) - Mt. Everest.

68-69 ● Khumbu (Nepal)
- Mt. Everest.

70-71 ● Khumbu (Nepal) -
Everest (left) and Nuptse
(right).

72-73 ● Khumbu
(Nepal) - Lhotse.

74 • Nepal - Gangapurna.

75 • Tibet (China) - Kailash.

Nepal - Annapurna.

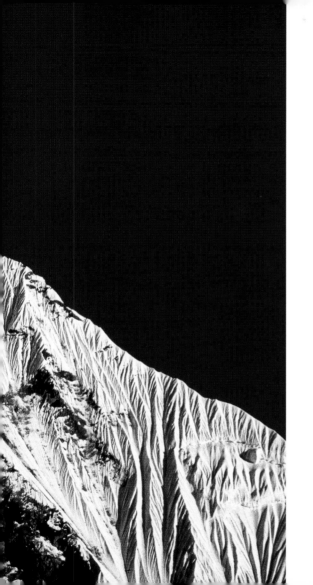

78-79 ● Nepal - Machapuchare,
Annapurna Sanctuary.

79 ● Karakorum (Pakistan) - Gasherbrun IV.

80-81 ● Nepal - Tent Peak, Annapurna
Sanctuary.

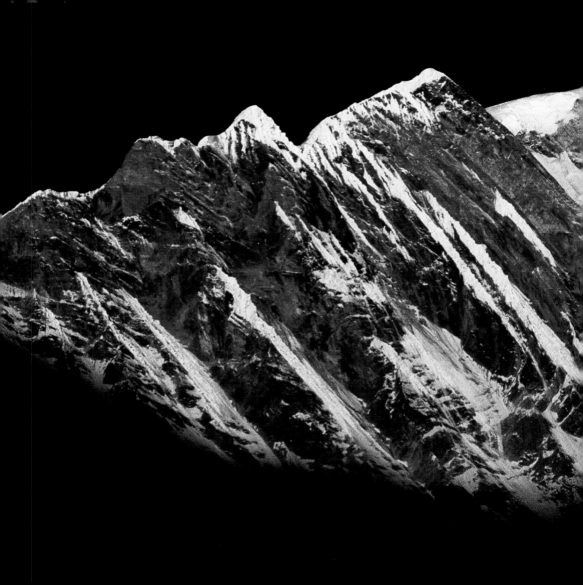

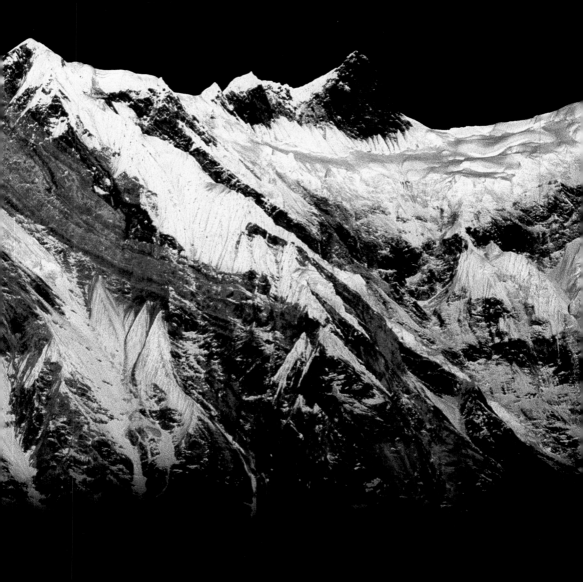

82 ● Uttar Pradesh (India) - Shivling Peak.

82-83 ● Uttar Pradesh (India) - Bagirathi Peaks.

84-85 ● Karakorum (Pakistan) - Masherbrun.

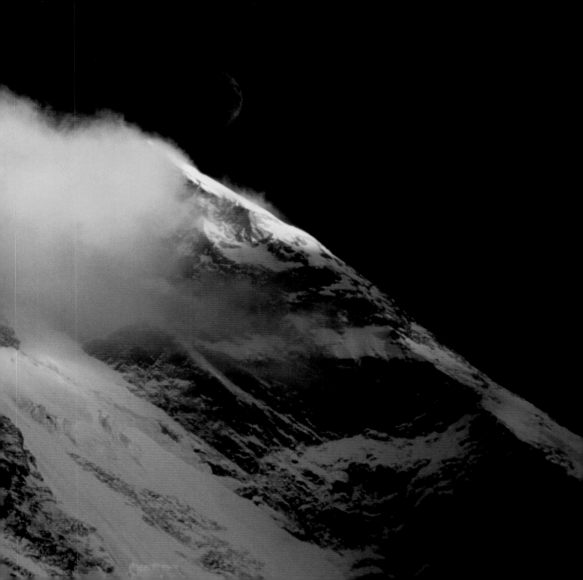

86-87 Karakorum
(Pakistan) - West face
of K2.

88-89 Yukon Territory
(Canada) - St. Elias
Mountains.

90-91 ● Alaska (USA) -
Mount McKinley.

92-93 ● Alaska (USA) -
Glacier Bay National Park
and Preserve.

94-95 ● Alaska (USA) -
Alaska Range.

96-97 • Alaska (USA) - Wrangell-St. Elias
National Park.

98-99 • Alberta (Canada) - Rampart Range
Peaks, Jasper National Park.

100-101 • Alberta (Canada) - Canadian
Rockies, Banff National Park.

102-103 • Wyoming (USA) - Grand Teton
National Park.

" THE SPIRIT OF A MOUNTAIN LIES IN ITS SNOWFIELDS AND CREVASSES, BEYOND THE LAST ROCK THAT EXTENDS TOWARDS THE SKY. ITS WORDS ARE WRITTEN IN THE CLOUDS THAT SHROUD THE SUMMIT AND ARE HEARD WHEN THE WIND HOWLS DOWN ITS PATHS, RECOUNTING THE STORIES OF THE MEN WHO HAVE TRUDGED THEM. "

104 ● Washington (USA) - Mt. Mount Olympus, Olympic National Park.

105 ● Washington (USA) - Mt. Mount Rainier.

106-107 ● California (USA) - Aerial view of Yosemite National Park.

108-109 ● California (USA) - Half Dome, Yosemite National Park.

110-111 ● California (USA) - Sequoia and Kings Canyon National Park.

112-113 ● Patagonia (Chile) - Torres del Paine.

114-115 ● Patagonia (Argentina) - The Fitz Roy Group, Parque Nacional los Glaciares.

116-117 ● Patagonia (Argentina) - Cerro Torre.

118-119 ● Peru - Huascaran, Cordillera Blanca.

120 ● Peru - Huandoy.

120-121 ● Peru - Huascaran,
Cordillera Blanca.

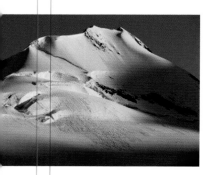

Antarctica - Fief Mountains,
Wiencke Island.

" THE SIGH OF THE WIND, YOUR FACE BITTEN BY THE COLD, BURNING LUNGS, LEGS LIKE BROKEN BRANCHES, TREMBLING HANDS AND A HOPELESSLY BLURRED VIEW. YET YOU GO ON, INCH BY INCH, PULLED TOWARDS THE TOP BY AN UNKNOWN AND INEXORABLE GOD. "

124-125 • South Island (New Zealand) - Southern Alps.

126-127 • South Island (New Zealand) - Mount Cook.

The BLUE DEPTHS

- Society Islands (French Polynesia) - Bora Bora.

INTRODUCTION The Blue Depths

SEEN FROM THE SKY, THE THIN SHEET THAT COVERS THE CONTINENTS IS A LUCID VEIL OF PURE SILK TORN HERE AND THERE; THESE RENTS REVEAL A SURFACE THAT IS WHITE, OR YELLOWISH, OR BRIGHT GREEN: DRY LAND. THIS MANTLE HAS MYRIAD NUANCES, ACCORDING TO THE DEPTH, SALINITY AND LATITUDE: GREEN AND LIGHT BLUE OFTEN MERGE WITH LIGHT AND REFLECTIONS, CREATING EXTRAORDINARY PLAYS OF TRANSPARENCY. SEEN FROM THE LAND, THE SEA IS A WAVE THAT CRASHES AGAINST THE CLIFFS, THE WHITE FOAM THAT SLIDES ALONG THE SHORELINE, THE LIGHT OF THE SETTING SUN OR THE MOONLIGHT THAT PAINTS THE SURFACE OF THE SEA AND THEN DIES OUT ON THE SHORE. IT IS A FISHING BOAT UNDER WAY TOWARD THE

INTRODUCTION The Blue Depths

OPEN SEA, LADEN WITH FISHERMEN AND PRAYERS, A SAILING BOAT ON THE HORIZON FOLLOWED BY SEA-GULLS WHOSE UNFURLED SAIL IS FILLED OUT BY THE WIND, A WAVING PALM TREE IN A DISTANT TROPICAL LOCALE. SEEN FROM THE DEPTHS, THE SEA IS A MYSTE-RIOUS WORLD FILLED WITH FANTASTIC SIGHTS. THERE ARE NIMBLE PREDATORS WITH THEIR GLASSY LOOK; SINUOUS, MULTICOLORED NUDIBRANCHIAN MOLLUSKS; FISH OF EVERY SHAPE AND COLOR; AND GIGANTIC CREATURES WITH HUGE APPETITES, AS WELL AS CAR-CASSES OF SHIPWRECKS ENCRUSTED WITH CORAL AND IMPRESSIVE CANYONS AND ABYSSES OUT OF WHICH THERE DART FLUORESCENT ECTOPLASMS, LIKE GHOSTS FROM THE HEREAFTER. FOR MAN, THE SEA

The Blue Depths
Introduction

MEANS THE SCENT OF BRINE AND THE SALT DEPOSITED ON ONE'S DOOR; IT IS A SHORT-LIVED LOVE STORY, THE LONGING FOR A LONG WALK ALONG THE BEACH, AN UNFORGETTABLE SONG, HOPE FOR OPPORTUNITIES AND A BETTER FUTURE. THE SEA IS ADVENTURE AND TALES OF HEROISM, IT IS A SAILING VESSEL MOVING OUT INTO THE OPEN WATERS, A SEAGULL FLYING OVER A SHIP THAT HAS BEEN IN THE OPEN OCEAN FOR MONTHS; IT IS THE STRIP OF LAND THAT APPEARS ON THE HORIZON, A HOLD FILLED WITH TREASURES FROM NEW WORLDS, THE BORDERLINE BETWEEN PAST AND FUTURE, BETWEEN A WASTED LIFE AND A LIFE LIVED TO THE FULL.

133 ● Sardinia (Italy) - Pink Beach, Budelli.

134-135 ● Micronesia - Foreshore, Lekes Sandspit, Palau.

136-137 ● Micronesia - Caroline Islands, Palau.

138-139 ● Marquesas Islands (French Polynesia) - Nuku Hiva.

140-141 ● Society Islands (French Polynesia) - Aerial view of Bora Bora.

142-143 ● Society Islands (French Polynesia) - Spectacular light on the reef, Bora Bora.

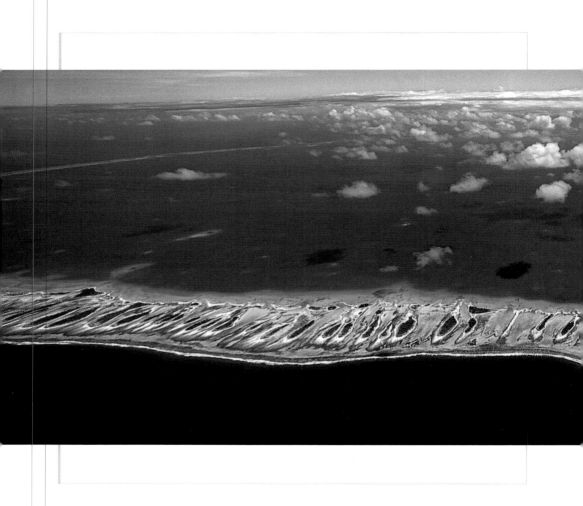

Tuamotu (French Polynesia) - Rangiroa Atoll.

146-147 ● Tuamotu (French Polynesia) - Island and channel on Rangiroa.

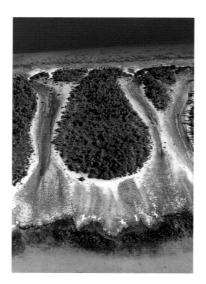

148-149 ● Tuamotu (French Polynesia) - Mataiva.
150-153 ● Society Islands (French Polynesia) - Effects of light
in the water at Huahiné.
154-155 ● Victoria (Australia) - The coast along the Great Ocean Road.

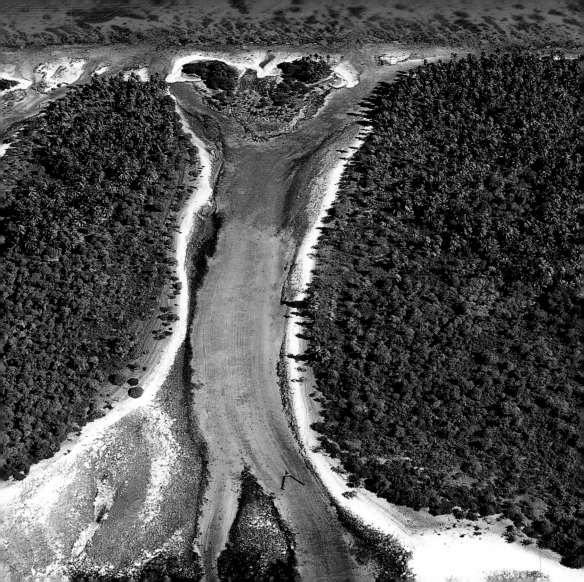

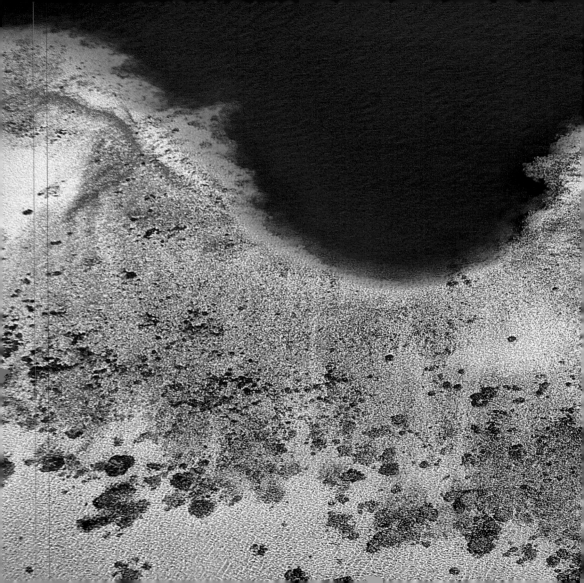

Queensland (Australia) - Whitsunday Island, Great Barrier Reef Marine Park.

158 ● Queensland (Australia) - Boult Reef, Capricorn Group.

159 ● Queensland (Australia) - Lady Musgrave Island, Capricorn Group.

160-161 ● Seychelles - La Digue.

162-163 Philippines - Islet off Palawan.

164-165 Thailand - Rai Lai beach, Krabi.

166-167 ● Red Sea (Egypt) - Gobal Strait.

168-169 ● Red Sea (Egypt) - Island of Tiran.

170-171 ● Tanzania - Beach near Bagamoyo.

172-173 ◆ Tanzania - Pacific
Ocean near Dar es Salaam.

174-175 ◈ Namibia - Skeleton
Coast.

176-177 ◈ Namibia - Bakers Bay.

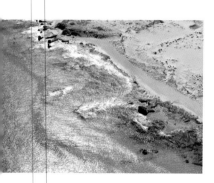

178-179 • Egypt - Mediterranean coast near Alexandria.

180-181 • Greece - Elafonisi, Crete.

182-183 • Sicily (Italy) - Rabbit Beach, Lampedusa.

184-185 • Sardinia (Italy) - Maddalena Archipelago.

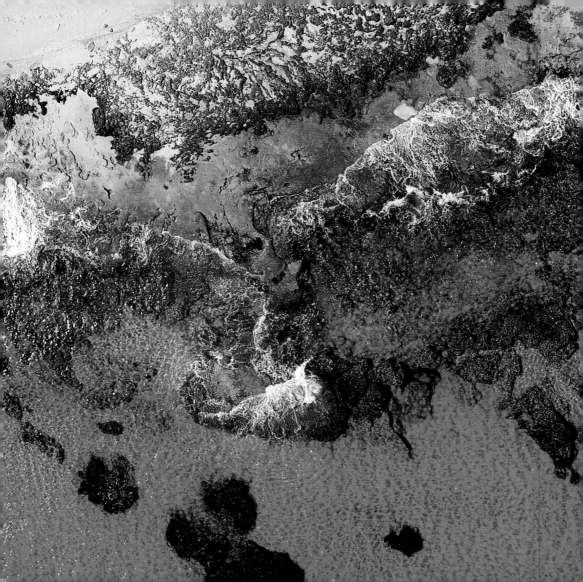

186 ● Puglia (Italy) - Pizzomunno Stack.

187 ● Puglia (Italy) - Zagare Cove.

188-189 ● Pontine Islands (Italy) -
Chiaia di Luna, Ponza.

189 ● Lipari Islands (Italy) -
The island of Vulcano.

190-191 ● Tuscany (Italy) -
The island of Capraia.

192-193 ● Balearic Islands (Spain) -
Conejera Island.

194-195 ● Algarve (Portugal) -
Coastline of Sagres.

195 ● Algarve (Portugal) - Praia
da Rocha.

196-197 ● Corsica (France) -
Southwest coast.

197 ● Côte d'Azur (France) - Petit
Langoustier, Porquerolles.

198-199 ● Brittany (France) -
Coastline near Finistère.

200 ● Normandy (France) - Coastline of the Pays de Caux.

200-201 ● Normandy (France) - Aval lighthouse, Étretat Bay.

202-203 ● Wales (United Kingdom) - Gower coast, West Glamorgan.

204-205 ● England (United Kingdom) - Bishop Rock in the English Channel.

206-207 ● County Sligo (Ireland) - Foreshore in Sligo Bay.

208-209 ● County Sligo (Ireland) -
Sligo Bay.

210-211 ● Cuba - Cayo Coco.

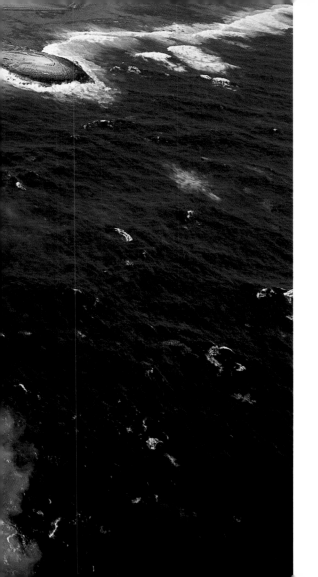

212-213 ● Guadeloupe (French Overseas Territories) - Pointe des Chateau.

214-215 ● Bahamas - Exuma Cays.

216-217 ● Colima (Mexico) - Manzanillo.

218-219 ● Yucatán (Mexico) - Las Coloradas.

220-221 ● California (USA) - Big Sur.

222-223 ● Big Island (Hawaii) -
Waipio Bay.

224-225 ● Kauai (Hawaii) - Napali Coast.

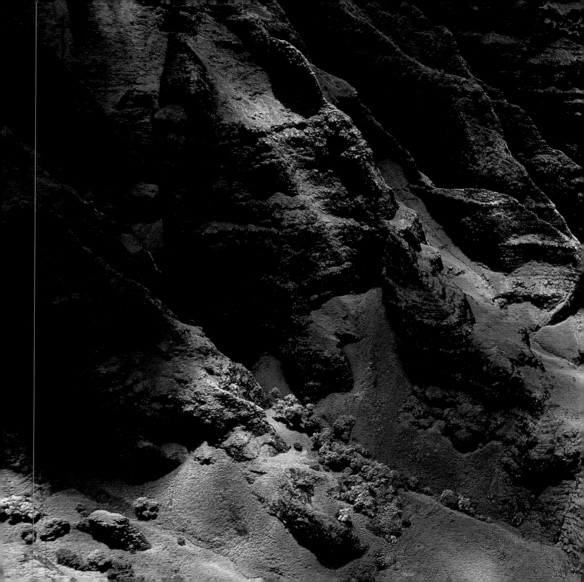

SILENT PLANET

● Sinai (Egypt) - Ain Um Ahkmed oasis.

INTRODUCTION Silent Planet

Sahara, Kalahari and Gobi are synonyms for lands of no return, journeys without hope, endless wasteland, unquenchable thirst and death. And yet, the eternal expanses of the world's largest deserts offer a unique fascination that requires stamina and sensibility, strength and patience to be appreciated. The most barren areas in the world do not forgive the superficial visitor: we are at the end of the world and the last sign of western civilization disappearing behind the earth-colored hills are the four-wheel-drive vehicles that cautiously venture along pebbly tracks. The few road signs that have survived the sandstorms are like bot-

INTRODUCTION Silent Planet

TLES THROWN INTO THE OCEAN BY A CASTAWAY: MAR-RAKESH 5,000 KILOMETERS, KUFRA OASES 2,500 KILOME-TERS, AIN-BEN-TILI 3,200 KILOMETERS. DESERTS ARE LIKE THE LIGHT THAT FLOODS THEM, CREATING UTTER CON-TRASTS: SUFFOCATING HEAT BY DAY AND BITTER COLD AT NIGHT; LONG TREKS INTO WASTELAND AND SUDDEN BURSTS OF LIFE; FIERY REDS, BRIGHT YELLOWS, SKY-BLUES, AS WELL AS THE TYPICAL COLORS OF ASH AND STONE. IN THIS DOMAIN OF BARRENNESS EVERYTHING SEEMS TO BE IN PERPETUAL MOVEMENT: WIND-SWEPT DUNES FOLLOW ONE ANOTHER, FORMING INCREDIBLE PLAYS OF SHADOWS; ROCKS SHAPED BY EROSION ON DISPLAY LIKE BIZARRE SCULPTURES AND ENCHANTED CASTLES; EVIL SPIRITS THAT CREATE AN ENDLESS MIRAGES TO MOCK THIRSTY WAYFARERS; EVEN THE

Silent Planet

Introduction

SOUND CREATED BY THE WIND AS IT SHIFTS THE GRAINS OF SAND IN INFINITE MODULATIONS. IN THESE DREAM-LIKE LANDSCAPES, SUBJECT TO CONTINUOUS, UNPREDICTABLE CHANGES, THERE LIVE ELUSIVE, OFTEN MYSTERIOUS CREATURES ABLE TO EXPLOIT SINGLE DROPS OF WATER, LIVING IN A SORT OF SUSPENDED STATE INTERRUPTED BY RARE MOMENTS OF VIGOROUS ACTION. AND THE DESERT PEOPLE ARE EQUALLY MYSTERIOUS, THEIR SKIN AS HARD AS THE LAND ITSELF. REMOTE AND INTANGIBLE, AS ABSOLUTE AS SPACE BUT, LIKE HUMANS, LACKING CER- TAINTY, DESERTS KINDLE OUR IMAGINATION WITH THEIR HARSH AND SPECTACULAR SCENERY. BUT THEY ARE ALSO A WORLD IN WHICH PARADOX BECOMES POETRY.

231 ● New South Wales (Australia) - 'Wall of China,' Mungo National Park.

232-239 ● Libya - Variations in the light and shapes of the dunes in the Libyan Desert.

240-241 ● Region of Fezzan (Libya) - Animal-shaped rocks near Akakus.

242-243 ● Region of Fezzan (Libya) - Sebha oasis.

244-245 ● Namibia - Large dune area in the Namibian Desert.

246 ● Namibia - Herd of ostriches
by a large dune.

246-247 ● Namibia - Light and shadow
in the dunes.

248-249 ● Namibia - Traces of erosion in
the desert to the west of Gamsberg Pass.

250-251 ● Namibia -
Sussusvlei dunes.

252-253 ● Namibia -
Clouds over the
Namibian Desert.

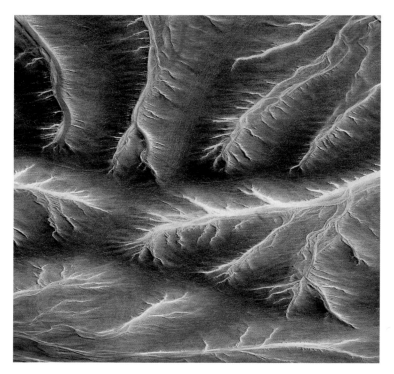

254-255 ● Egypt - Branching of a wadi in the desert north of Bahariya.

256-257 ● Egypt - Wind-sculpted dunes near Bahariya and Farafra oases.

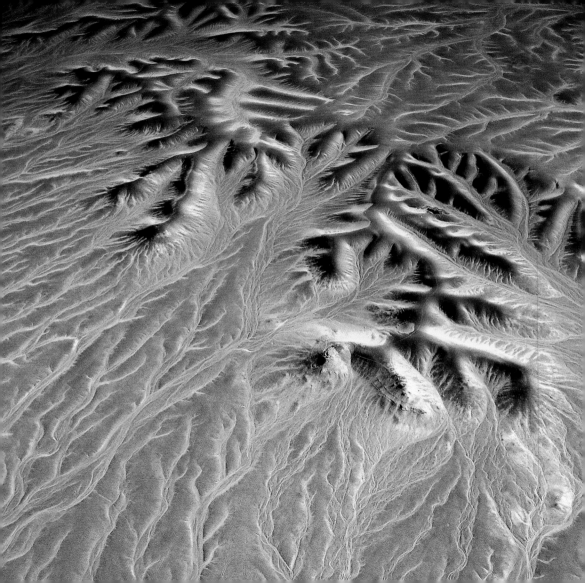

258-259 ● Egypt - White Desert.

260-261 ● Egypt - White Desert near Farafra.

262-263 ● Egypt - Siwa Oasis.

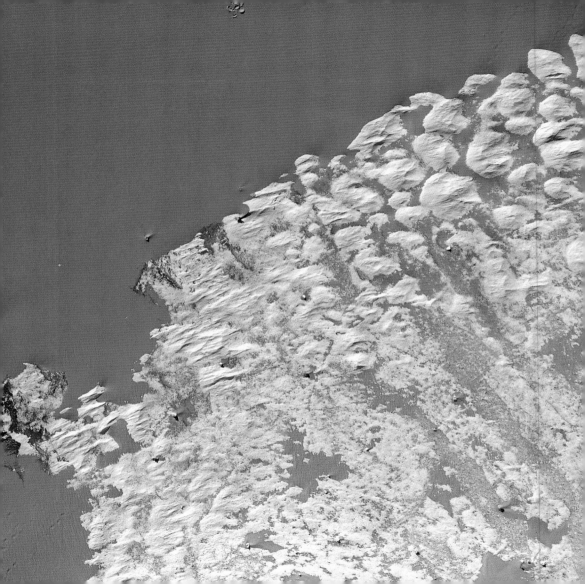

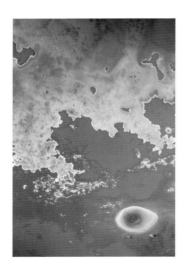

Egypt - Salt lakes near Shali,
Siwa Oasis.

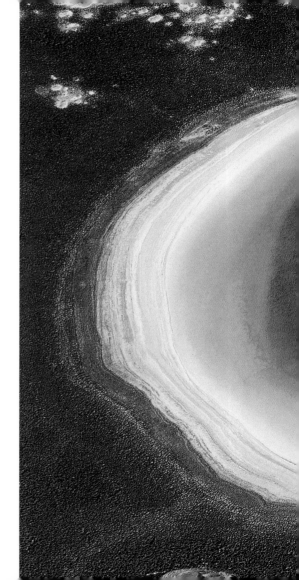

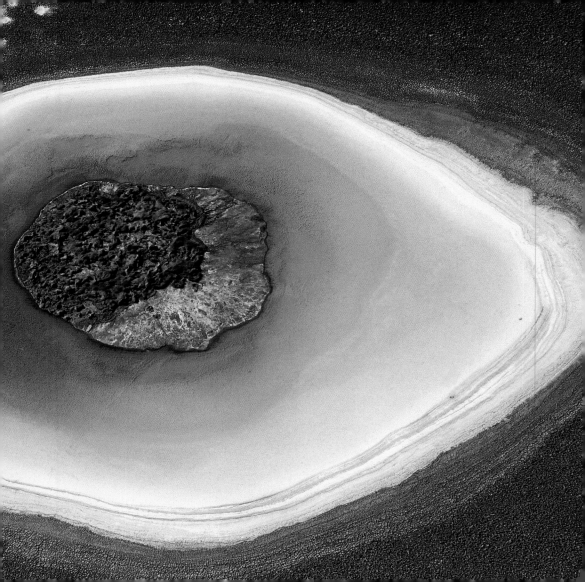

"THE COLORS, WHICH RANGE FROM GOLD THROUGH ORANGE AND OCHER TO BURNT RED, ARE THE UNIFYING ELEMENT IN THE SAND DESERTS OF THE WORLD. FAR FROM THE MONOTONY OF DESOLATE STONEY LANDS, THE DUNES OF THE SAHARA AND GOBI DESERTS OFFER A PALETTE OF MANY COLORS TO THE INCREDULOUS EYES OF THE OBSERVER. "

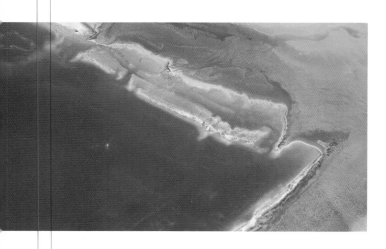

266-267 ● Egypt - Mineral lakes near Fayum Oasis in the Eastern Sahara.

268-269 ● Sinai (Egypt) - Jebel Musa.

270-271 ● Israel - Negev Desert.

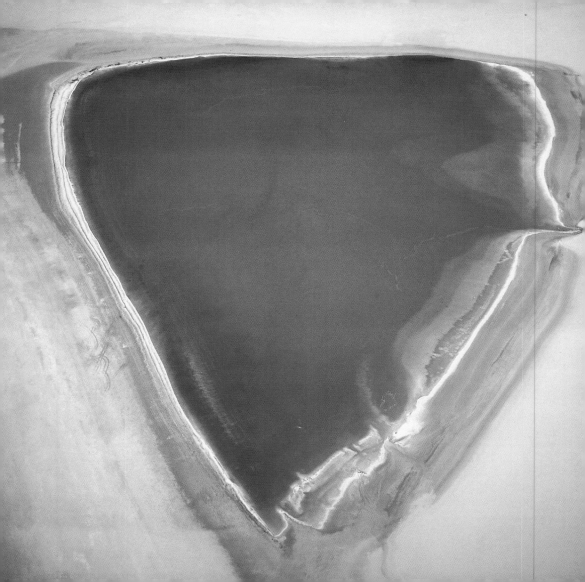

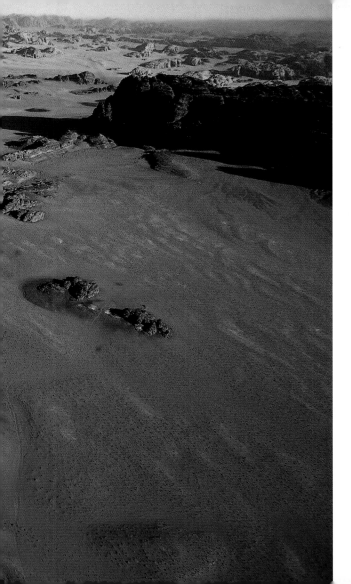

272-273 ● Jordan - Wadi Rum.

274-275 ● California (USA) - Death Valley, Zabriskie Point.

276-277 ● California (USA) - Death Valley, Badwater Pool.

278 ● Arizona (USA) - Superstition Mountains.

278-279 ● Arizona (USA) - Sonora Desert.

280-281 ● Arizona (USA) - Colorado River Valley.

282-283 ● Xinjiang Uygur (China) - Erosion-formed valley on the Silk Road.

284 ● South Australia (Australia) - Strzelecki Desert.

285 ● South Australia (Australia) - Everard Ranges, Great Victoria Desert.

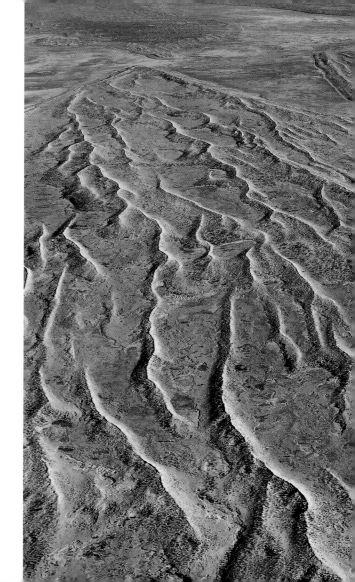

286-287 ◆ South
Australia (Australia) -
Simpson Desert.

288-289 ◆ Western
Australia (Australia) - The
Pinnacles, Nambung
National Park.

ROOTS
of the
WORLD

Nigeria - Jungle in the Niger Delta.

INTRODUCTION Roots of the World

THE "LUNGS" OF THE EARTH ARE GREEN. A BRILLIANT GREEN LIKE MOUNTAIN WOODS, EMERALD LIKE THE SOUTH AMERICAN PLUVIAL FORESTS, TINGED WITH BLUISH NUANCES LIKE THE CONIFERS IN NORTH EUROPE, BRIGHT LIKE THE SUNNY PATCHES IN THE SOUTH MEDITERRANEAN OR THE DATE PALMS SURROUNDED BY SAND IN NORTH AFRICA. VIEWED FROM THE SKY, FORESTS ARE COLORED CLOUDS, AS THICK AS CARPETS. SEEN FROM GROUND LEVEL, THEY ARE THE LAND OF FANTASY, DREAMS, LIFE ITSELF. IN FACT, FORESTS HAVE ALWAYS BEEN THE SETTING FOR THE MOST FABULOUS FAIRY TALES AND SAGAS; THEY ARE THE HOME, THE WOMB, THE PLACE OFFERING WOOD AND FOOD TO KEEP US WARM AND FEED US, AND SHELTER US FROM RAIN OR HEAT.

INTRODUCTION Roots of the World

SOMETIMES FORESTS ARE ALSO PLACES OF THE SPIRIT, AS FOR EXAMPLE IN DANTE, WHO VIEWED THE "DARK WOOD" AS THE ABODE OF SIN, TERROR AND EVIL, WHICH ASSAIL HUMANS DURING THEIR JOURNEY TOWARD LIGHT. THIS IS ALSO THE CASE WITH TOLKIEN, WHOSE SWEETEST, MOST INTELLIGENT, STRONGEST AND HEROIC CREATURES LIVE IN THE WOODS, WHILE EVIL AND VIOLENCE DWELL IN FLAT STEPPES, ON OMINOUS MOUNTAINTOPS OR IN DARK CAVES. FORESTS LIVE IN SYMBIOSIS WITH WATER. THE BEST EXAMPLES OF THIS ARE TROPICAL FORESTS, WHERE IN CERTAIN SEASONS EVERY CREEK BECOMES A TORRENT AND EVERY TORRENT A FORMIDABLE RIVER THAT CARRIES TO THE SEA TONS OF SEDIMENT AND TREE TRUNKS RIPPED AWAY FROM THE BANKS, AND

Roots of the World
Introduction

EQUATORIAL JUNGLES, CALLED "CLOUD FORESTS" BECAUSE THEY ARE ALMOST ALWAYS IMMERSED IN THICK RAIN CLOUDS. THE SAME SYMBIOSIS EXISTS BETWEEN WOODS AND THE MYRIAD CREATURES THAT LIVE IN THEM: FLOWERS, BUTTERFLIES, HYMENOPTERA AND COLEOPTERA WITH THEIR DAZZLING COLORS, FISH, REPTILES, AMPHIBIANS, BIRDS, MAMMALS LARGE AND SMALL. A SELF-SUFFICIENT MICROCOSM; A BIOSPHERE IN WHICH THE CHAIN OF LIFE HAS A THOUSAND BEGINNINGS AND A THOUSAND CONCLUSIONS; AN ENVIRONMENT BUFFETED BY LIGHT WHEN IT FILTERS THROUGH THE FOLIAGE IN BROAD SHAFTS, BUT WHICH BECOMES MORE FASCINATING IN DARKNESS AND MYSTERY.

295 ● Quebec (Canada) - Forest in fall.

296-297 ● Piedmont (Italy) - Larch woods in Valsesia.

298-299 • Lapland (Sweden) - Forest in winter.

300-301 ● Friuli Venezia Giulia (Italy) - Forest in Tarvisio.

302-303 ● Tuscany (Italy) - Cluster pines in Maremma.

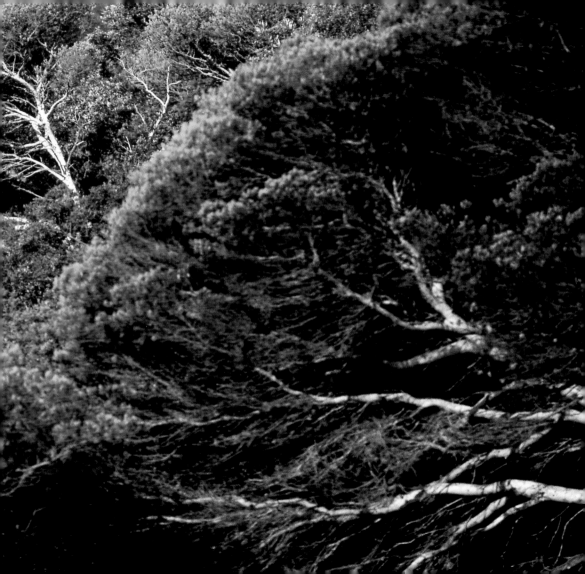

304-305 ● Bavaria (Germany) - Ferns and clover in a Bavarian forest.

306-307 ● Denmark - Nordskoven Forest.

308-309 ● Nigeria - Jungle in the Niger Delta.

310-311 ● Congo - Forest along the Konilou River.

312-313 • Congo - Mayombe
Forest.

314-315 • Congo - Forest
of bamboo.

316-317 • Congo - 'Gorges de
Diosso' at Pointe Noire.

318-319 • Tanzania - Lake
Victoria, Rubondo Island.

320-321 • Colorado (USA) - Forest of poplars and fir in Aspen.

322-323 • Colorado (USA) - Banded Peak.

324-325 • Colorado (USA) - Forest in Aspen during an Indian summer.

326-327 • California (USA) - Redwood National Park.

328-329 • California (USA) - Sequoias in Redwood National Park.

330-333 ● Florida (USA) -
Tropical vegetation in
Lake Woodruff National
Wildlife Refuge.

334-335 ● Costarica -
Monteverde Forest.

336-337 ● Costarica -
Tortuguero National Park.

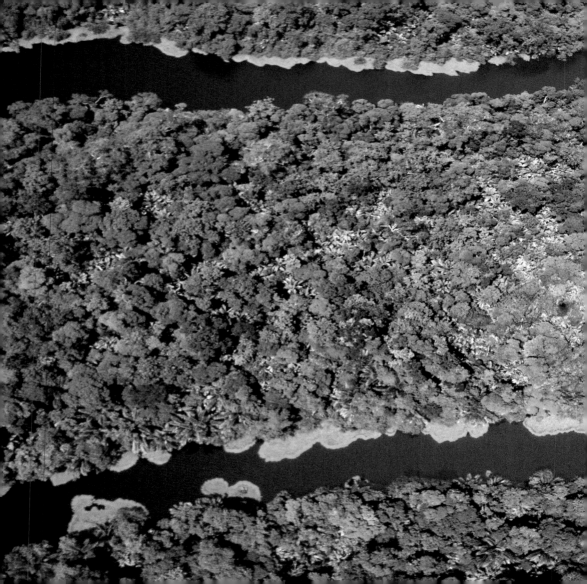

338-339 ● Sri Lanka - The flower
of *Gloriosa superba*,
Udowattakele Forest Reserve.

339 ● Sumatra (Indonesia) -
Balanophora, Gunung Leuser
National Park.

340-341 ● Borneo (Malaysia) -
Sabah Forest.

342-343 ● Borneo (Malaysia) -
Kinabalu National Park.

344-345 ● Society Islands
(French Polynesia) -
Huahiné Forest.

346-347 ● Tuamotu Islands
(French Polynesia) - Palm trees
on Rangiroa Atoll.

Victoria (Australia) - East Gippsland
Errinundra National Park.

350 • Tasmania (Australia) - Mount Field National Park.

351 • Tasmania (Australia) - Franklin L. Gordon Wild Rivers National Park.

352-353 • Tasmania (Australia) - Pelorus River Scenic Reserve.

354-355 • Coromandel Peninsula (New Zealand) - Coromandel State Forest Park.

A SENSE of WATER

Canaima (Venezuela) - Angel Falls.

INTRODUCTION A Sense of Water

IF RIVERS ARE AN ALLEGORY OF LIFE, WATER-FALLS ARE THE MOMENTS IN WHICH LIFE ACCELERATES AND PRECIPITATES, LEAVING US BREATHLESS. STILLNESS AND ROARING, FERTILE STRENGTH AND PURE ENERGY, SLOW TEMPO AND PRESSING RHYTHM …. EVERY WATERWAY EXPRESSES THE CONTRADICTIONS OF EXISTENCE AND IS AT ONCE TENDER AND VIOLENT, PLACID AND ANXIOUS. HOW DIFFERENT THE PARANÁ RIVER IS AS IT TRAVERSES ARGENTINA SO MEEKLY AND THEN BECOMES FEARSOME AS IT PLUNGES OVER THE IGUAZÚ FALLS. HOW DIFFERENT TOO IS THE ST. LAWRENCE RIVER, WITH NIAGARA FALLS AND LAKE ONTARIO TO THE WEST, FLOWING UNHURRIEDLY TO ITS VAST, FRIGID ESTUARY IN THE ATLANTIC OCEAN. A RIVER IS ABUNDANCE

INTRODUCTION A Sense of Water

AND ARTISTIC BEAUTY. IT IS A PAINTING BY AN 18TH-CEN-
TURY LANDSCAPIST WHO METICULOUSLY DEPICTS THE
WEEPING WILLOWS WHOSE BRANCHES ALMOST TOUCH
THE WATER; CLUSTERS OF LONG, STRINGY GRASS THAT
FLUTTER LIKE A FAIRY'S HAIR IN THE CURRENT; WAVES
THAT SUDDENLY ACCELERATE BECAUSE OF BENDS OR
NARROWS IN THE RIVER AND ARE STREAKED WITH WHITE
FOAM; THE CRYSTAL-CLEAR WATER CARESSING THE
ROCKS .... DOWNSTREAM, WHERE THE WATER FLOWS
MORE SLOWLY, RIVERS BECOME NAÏF PAINTINGS WITH
THE MURKY COLORS OF THE AMAZON RIVER. HERE
THERE ARE NO LONGER THE DARK GREENS OF THE
BANKS, THE LEADEN HUES OF THE DEPTHS, THE FLICK-
ERING REFLECTIONS OF THE MANY RUNNELS CREATED

INTRODUCTION A Sense of Water

WHEN THE CURRENT BECOMES A BURST OF ELECTRIC ENERGY. THE DOMINATING SHADES ARE NOW BROWN, OCHER AND GRAY, THE LIGHT BECOMES SOFT AND THE AIR IS MORE HUMID. DRIFTS OF FOG SLIDE SLOWLY ALONG THE QUIESCENT WATER, FROM WHICH EMERGE REEDS AND HYDROPHILOUS PLANTS; WE CAN IMAGINE THEIR SUBDUED MUFFLING DROWNED OUT BY THE CRY OF THE BIRDS BICKERING ON EITHER SIDE OF THE RIVER. AT THE ESTUARY THE ABRUPT MIXTURE OF SALT AND FRESH WATER MAKES THE WATER BRACKISH, AND THE SCENT OF ALGAE MINGLES WITH THE STAGNANT RIVER AIR. WATERFALLS ARE THE LYRIC OF A ROMANTIC COM-POSER WHO IMAGINES MAN AWE-INSPIRED BY THE FORCE OF NATURE. FIRST THERE IS A CONTINUOUS AND

INTRODUCTION A Sense of Water

IMPRESSIVE TREMOR, THEN A SUBDUED RUMBLE COV-
ERED BY THE SOUND OF THE WIND AMONG THE FOLIAGE
AND BIRDS SINGING, AND LASTLY A MAJESTIC ROAR AND
GUSTS OF WIND. AFTER WHICH THERE IS ONLY THE
ENCHANTED HARMONY OF GENERATING POWER IN ITS
PUREST FORM, A FABULOUS AND TERRIFYING SPECTA-
CLE OF VAPORIZED DROPS OF WATER THAT CREATE
THOUSANDS OF RAINBOWS, SWEET VISIONS THAT CON-
TRADICT THE HURRICANE-LIKE VIOLENCE OF THE FOAM
ON THE ROCKS BELOW. IN THIS RESPECT, WATERFALLS
ARE FRAGMENTS OF A LOST PARADISE, PERHAPS OF A
LAND OF WHICH WE STILL HAVE ANCESTRAL MEMORIES.

362-363 ● Friuli Venezia Giulia (Italy) - Fusine River.

364-365 ● Saxony (Germany) - The Elbe River.

Norway - The
Alattiojoki River at
Kautokeino.

Finland - Oulanka
Joki River, Oulanka
National Park.

Burgundy (France) -
Gironde estuary.

County Donegal (Ireland) - Mouth of the Gweebarra River.

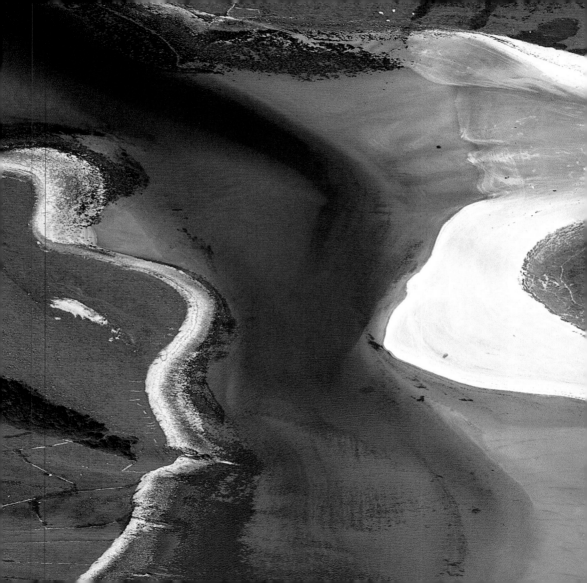

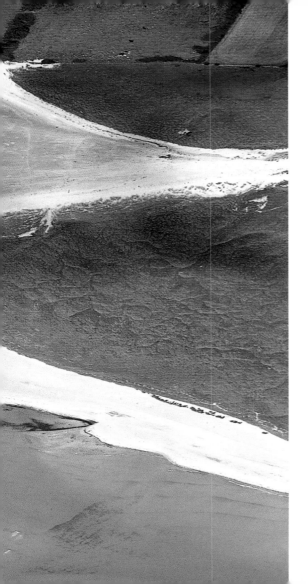

County Donegal (Ireland) -
Mouth of the Gweebarra River.

Rwande - The Kagera.

378-379 ● Tanzania - The Rufiji.

379 ● Nigeria - The Niger.

380-381 ● Zimbabwe - Victoria Falls.

382-383 ● Congo - The Kouilou.

384-385 ● Congo - Loufoulakari Falls.

386-387 ● Congo - Rapids in the River Congo.

388-389 ● Egypt - The Nile.

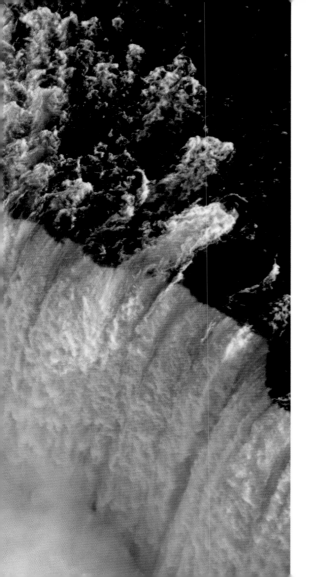

- Ontario (Canada) - Niagara Falls.

392 ● Northwest Territories (Canada) - Peace River Delta.

392-393 ● Alberta (Canada) - Slave River.

Wyoming-Montana-Idaho (USA) -
Yellowstone River.

Arizona (USA) - Grand Canyon with the Colorado River, Yavapai Point area

398-399 ● Molokai (Hawaii, USA) - Waterfall in Kalaupapa National Historical Park.

399 ● Kauai (Hawaii) - Waterfalls on the sides of the Waialeale volcano.

400-401 ● Brazil - Rio Negro and Arcipelago Anavilhanas.

402-403 ● Paranà, (Brazil) - Iguazú Falls.

404-405 ● Peru - The Tigre.

406-407 ● Venezuela - Sand banks on the Orinoco.

408 ● South Island (New Zealand) -
Sutherlands Falls.

408-409 ● South Island (New Zealand) -
Bowen Falls.

410-411 ● Papua New Guinea - Early
morning mist on the Karawari River.

412-413 ● Tibet (China) - The Brahmaputra Valley.

414-415 ● Tibet (China) - The wide bed of the Brahmaputra.

The MIRRORS of the SKY

● Arizona (USA) - Lake Powell.

INTRODUCTION The Mirrors of the Sky

IN TWILIGHT IN ITALY, D.H. LAWRENCE DESCRIBES
AN IRIDESCENT VAPOR THAT ROSE FROM THE WATER
TOWARD THE VILLAGES OF LAKE GARDA, WHICH WERE
LIKE WHITE DOTS IN THE DISTANCE. EMERGING FROM THE
TUFTS OF FOG OVER THE SURFACE OF THE LAKE WERE
BOATS WITH ORANGE SAILS GLIDING OVER THE
TURQUOISE WATER. A SIGHT OF SHEER BEAUTY, "LIKE PAR-
ADISE, LIKE THE FIRST CREATION." LAKES ARE ALMOST
ALWAYS SYNONYMOUS WITH PEACE AND QUIET. THE TINY
SHEETS OF WATER IN THE ALPS ARE LIKE BLUE DROPS
SLIDING OVER THE GREEN VELVET OF THE PINE GROVES.
THE LARGE MORAINAL AND GLACIAL BASINS THAT APPEAR
IN PLAINS LIKE MINIATURE FRESHWATER SEAS: VOLCANIC
DEPRESSIONS FED BY RAINWATER OR BY GENEROUS

INTRODUCTION The Mirrors of the Sky

SPRINGS THAT FORM ROUND NATURAL POOLS. AND THEN THE MICROSCOPIC GREEN LAKES THAT POP UP FORM THE THICK VEGETATION OF TROPICAL ISLANDS, THE NUMEROUS LOW EXPANSES OF WATER THAT CROP UP IN THE FORESTS OF FINLAND, THE HIGH-ALTITUDE LAKES IN THE ANDES, THE LARGE AFRICAN BASINS THAT GIVE RISE TO THE NILE .... EACH WITH ITS OWN PARTICULAR SCENERY, CLIMATE, AND HISTORY. AN UNRECOGNIZED MULTITUDE, A HOST OF SILENT WATER MICROCOSMS, FRAGMENTS OF SKY THAT HAVE FALLEN TO THE EARTH WITH THEIR FASCINATION AND BEAUTY.

A FEW LAKES HAVE BECOME FAMOUS BECAUSE A DRAMATIC EVENT OCCURRED THERE, A MYSTERIOUS CREATURE IS SAID TO BE HIDING IN ITS DEPTHS, OR A NOVEL HAS

The Mirrors of the Sky
Introduction

BEEN SET ALONG ITS BANKS. MANY LESSER LAKES ARE KNOWN ONLY TO SHEPHERDS AND HIKERS. OTHERS SEEM TO HAVE BEEN CREATED BY THE DREAM OF A LOVING GOD, WITH THEIR TURQUOISE HUES, GOLDEN REFLECTIONS, THICK VEGETATION, MOUNTAINS MIRRORED IN THE WATER, AND EVERYTHING ELSE THAT MAKES FOR A PICTURE POSTCARD. YET OTHERS CONCEAL ARCHEOLOGICAL TREASURES OR HAVE TRIGGERED TENDER OR VIOLENT LEGENDS. IN ANY CASE, MOST LAKES ARE THE HOME OF ECOLOGICAL WONDERS THAT ALSO HOST ARTISTIC OR ARCHITECTURAL GEMS. THEY ARE BODIES OF WATER THAT FOR MILLENNIA HAVE REFLECTED THE SKY ABOVE AND THE EVENTS OF HUMAN HISTORY BELOW.

421 ● Yukon Territory (Canada) - Kluane Lake.

422-423 ● Quebec (Canada) - Laurentian Highlands.

424-425 ● Alberta (Canada) - Wedge Pond and Mount Kidd.

426-427 ● Maine (USA) - Moosehead Lake.

428-429 ● Yukon Territory (Canada) - Kluane Lake.

Northwest Territories (Canada) -
Great Slave Lake near
Yellowknife.

Wyoming (USA) - Lake Jackson, Grand
Teton National Park.

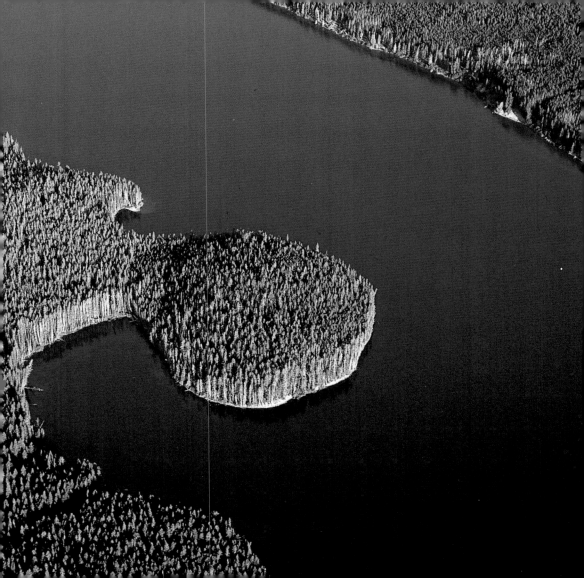

434 ● Alberta (Canada) - Moraine Lake, Banff National Park.

435 ● British Columbia (Canada) - O'Hara Lake, Yoho National Park.

436-437 ● Alberta (Canada) - Peyto Lake, Banff National Park.

438-439 ● Arizona (USA) - Lake Powell.

440-441 ● Chile - Lake Pehoe, Torres del Paine National Park.

442 ● Queensland (Australia) -
Lake Boomanjin, Fraser Island.

442-443 ● Queensland (Australia)
- Lake McKenzie, Fraser Island.

444-445 ● Afghanistan - Lake
Band i Amir.

446-447 ● Tibet (China) -
Lake Yamdruk.

Sichuan (China) - Lake Shuzheng.

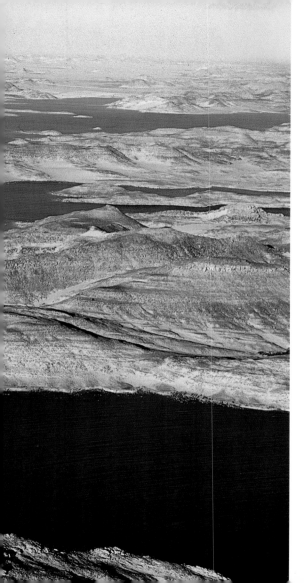

450-451 • Nubia (Egypt) - Lake Nasser.

452-453 • Egypt - Jebel Bayda and Lake Siwa.

454-455 ● Tanzania - Lake Victoria.

456-457 ● Tanzania - Lake in Serengeti
National Park.

458-459 ● Tanzania - Lake Natron.

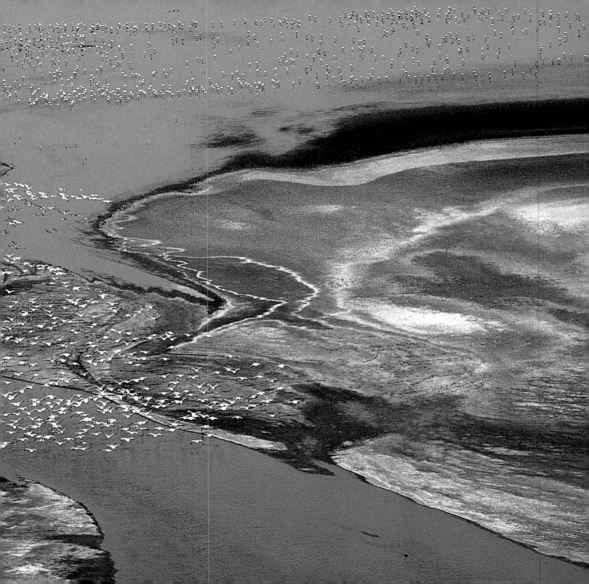

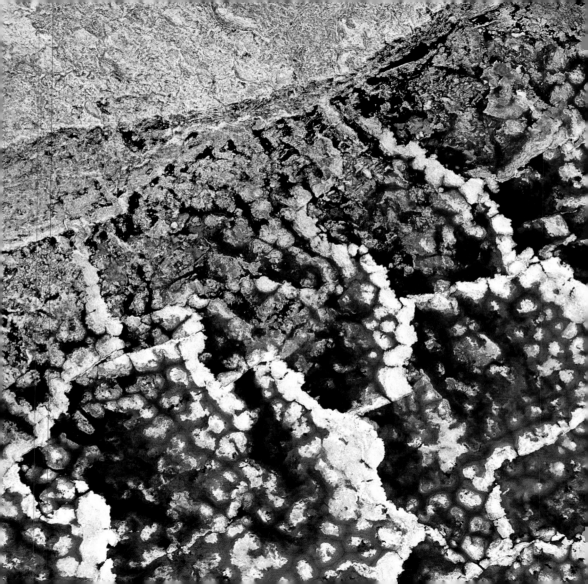

460-461 ● Tanzania - Patterns formed
by salt crystals on the banks of
Lake Natron.

462-463 ● Kenya -
Pink flamingoes, Lake Magadi.

464-465 ● Friuli Venezia Giulia (Italy) -
Fusine Lakes Natural Park.

466-467 ● England (United Kingdom) -
The lakes of Tarn Hows, Lake District,
Cumbria.

468-469 ● Norrland (Sweden) -
Marssjön near Marsliden.

The GODS of FIRE

Sicily (Italy) - Eruption of Etna.

INTRODUCTION The Gods of Fire

THE EARTH TREMBLING AND CRACKING OPEN, MOLTEN ROCKS EMERGING, EXPLOSIONS AND CLOUDS OF ASH, THUNDEROUS ROARS, RIVERS OF LAVA GLOWING AT NIGHT: VOLCANOES ARE SPECTACLES OF PURE VIOLENCE, A PRIMEVAL ECHO MIRRORING THE EARTH BOILING UNDER ITS CRUST. DUE TO THEIR OVERWHELMING EVOCATIVE NATURE AND EXTRAORDINARY DESTRUCTIVE POWER, VOLCANOES ARE THE SUBJECTS OF MANY ANCIENT MYTHS. FROM VESUVIUS TO KRAKATOA AND PELÉE, THOUSANDS OF PERSONS HAVE LOOKED THROUGH THE FALLING LAPILLI AT THE SKY, SEEKING AN EXPLANATION FOR THIS PHENOMENON. IN THE BEGINNING, GODS AND DEMONS CREATED THE "MOUNTAINS OF FIRE." FOR THE ANCIENT GREEKS, IT WAS THE TITANS WHO, IN THEIR STRUGGLE AGAINST THE

INTRODUCTION The Gods of Fire

OLYMPIC GODS, CAUSED THE ERUPTIONS AND SHOOK THE EARTH UNTIL FIRE SPRANG OUT OF THE VOLCANOES. THEIR INCANDESCENT BREATH FLOATED OUT OF THE CRATER, SPEWING MOLTEN ROCKS AND THEIR VIOLENT CRIES DEAFENED AND TERRIFIED PEOPLE. THE VOLCANO-GOD OF THE ANCIENT ISRAELITES WAS SINISTER, IRASCIBLE, NOISY, HIS VOICE RESOUNDING LIKE THUNDER AND OFTEN HERALDED BY A CLOUD, WHILE AT OTHER TIMES HE HIMSELF WAS A CLOUD. HE DID NOT LIVE ON MT. SINAI BUT MOVED FROM MOUNTAIN TO MOUNTAIN, MAKING THE EARTH TREMBLE AS HE WALKED, WHILE SMOKE CAME OUT OF HIS NOSTRILS AND FIRE SPURTED FROM HIS MOUTH. HE PUNISHED SINNERS BY PELTING THEM WITH FIRE AND BRIMSTONE. IN JAPAN, THE LAND OF VOLCANOES, IT IS SAID THAT CRATERS

The Gods of Fire
Introduction

ARE THE ABODE OF ONI, A SNEERING RED MONSTER WHO
THRASHES ABOUT AND HURLS STONES DURING ERUPTIONS.
IN THE HAWAIIAN ISLANDS THE CULT OF PELÉ, THE POLYNE-
SIAN GODDESS OF FIRE, IS STILL QUITE POPULAR. AFTER A VI-
OLENT ARGUMENT WITH HER SISTER, PELÉ WAS OBLIGED TO
SWIM TO THE SOUTH, AND EVERY TIME SHE EMERGED FROM
THE WATER AN ISLAND WAS BORN, THUS CREATING THE
HAWAIAN ARCHIPELAGO. PELÉ'S PRESENT HOME IS THE
HALEMAUMAU CRATER ON THE KILAUEA VOLCANO. WHEN
PELÉ GETS ANGRY, SHE STAMPS HER FOOT ON THE GROUND,
CAUSING THE CRUST TO BREAK AND LAVA TO SPEW OUT, A
"CAPRICIOUS" SPECTACLE THAT HAS AFFORDED SOME OF
THE MOST BEAUTIFUL PICTURES EVER TAKEN OF A VOLCANO.

475 ● Hawaii (USA) - Mouth of Kilauea, Hawaii Volcanoes National Park.

476-477 ● Washington (USA) - Mount St. Helens.

● Tanzania - Caldera on Kilimanjaro.

Tanzania - Crater of Oldonyo Lengai.

482 ● Vestmannaeyjar (Iceland) -
Eruption of Helgafell.

482-483 ● Iceland - Sub-glacial
eruption of Grimsvotn volcano.

484-485 ● Sicily (Italy) - Lava flow on Etna.

486-487 ● Hawaii - Lava flow on Kilauea, Hawaii Volcanoes National Park.

488-489 ● Hawaii - Eruption of Kilauea, Hawaii Volcanoes National Park.

＂ EARTH AND FIRE. POWER AND VIOLENCE. IMMENSE FORCES UNLEASHED BY IMPERCEPTIBLE MOVEMENTS BENEATH THE EARTH'S CRUST SEND MAGMA TOWARDS THE SURFACE. THE ERUPTION IS ONLY THE LAST ACT IN PHYSICAL ACTIVITY THAT HAS LASTED FOR MILLENNIA, AND WHICH REGARDS THE NATURE OF THE PLANET ITSELF. ＂

• Hawaii - Lava spout on Kilauea, Hawaii Volcanoes National Park.

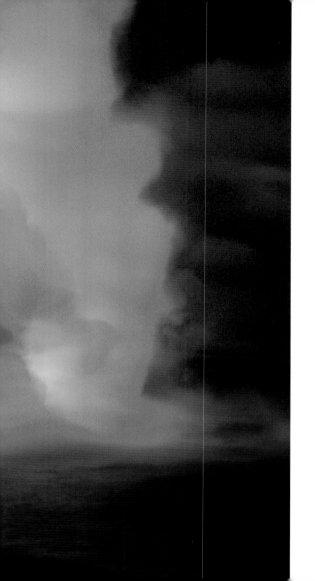

492-493 ● Hawaii - Lava flow from Kilauea crater, Hawaii Volcanoes National Park.

493 ● Hawaii - Lava spout on Mauna Ulu, Hawaii Volcanoes National Park.

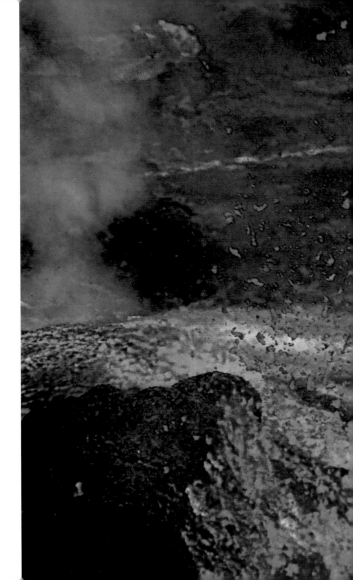

494-497 ● Île de la Réunion (French Overseas Territories) - Eruption of Piton de la Fournaise.

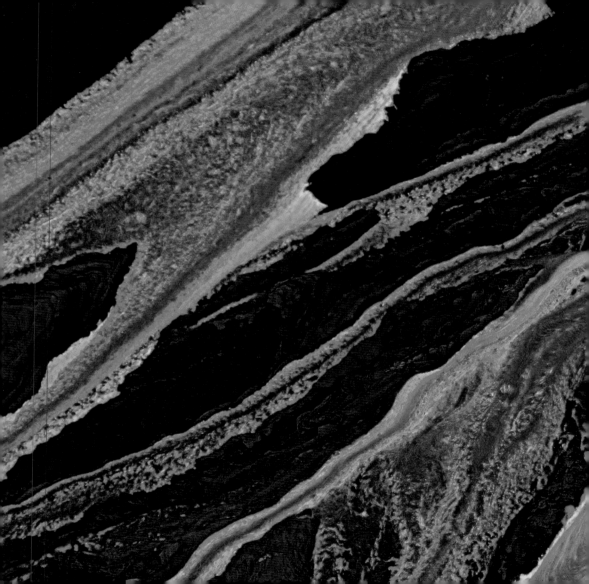

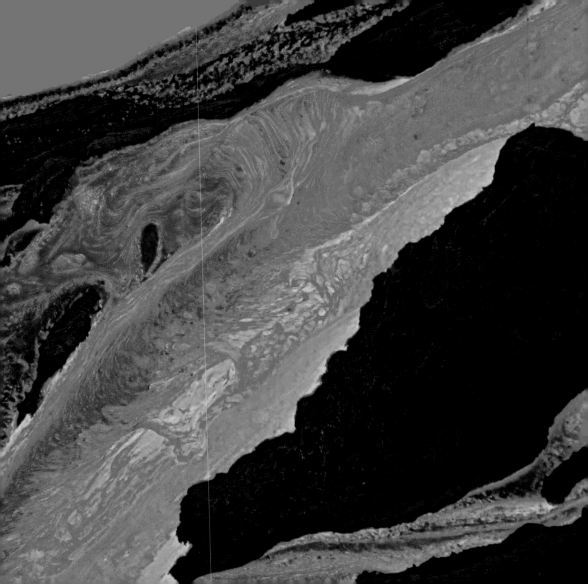

The ICE KINGDOM

Australian Antarctic Territory - Iceberg.

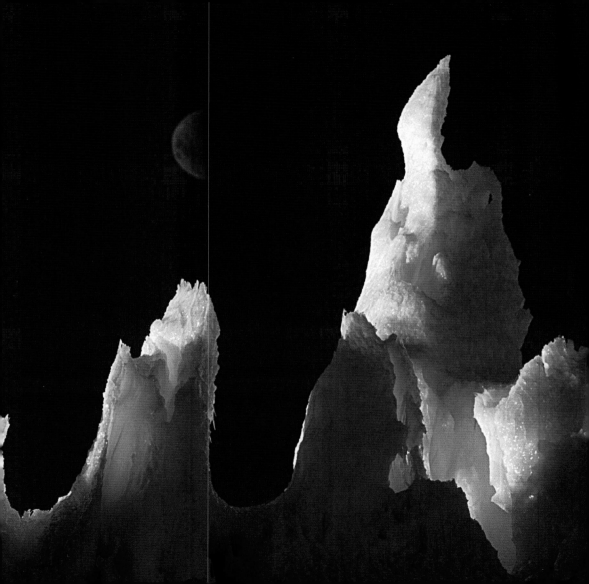

INTRODUCTION The Ice Kingdom

AMONG THE CRYSTAL TOWERS AND ALONG THE WIND-SWEPT PASSAGES IN THE WORLD OF GLACIERS, THE RULES OF LIFE LOSE THEIR MEANING: THERE REMAIN THE DIMENSIONS OF THE REAL AND IMAGINARY, WHICH MINGLE SOMEWHERE BETWEEN THE TRUE AND FALSE, WAKEFULNESS AND DREAM. TWO DIMENSIONS THAT ATTEST TO THE VAST WHITE EXPANSE THAT IS BLINDING ON SUNNY DAYS, MILKY ON CLOUDY DAYS, AND MURKY AND UNDECIPHERABLE UNDER THE MANTLE OF THICK, ICY FOG. THE THIRD DIMENSION IS A GIFT THAT CANNOT BE PERCEIVED AT THIS LATITUDE: HERE, WHERE THERE IS NO VEGETATION AND THE SOLID ICE IS ALWAYS COVERED WITH SNOW, THE BLINDNESS CAUSED BY THE LACK OF REFERENCE POINTS ANNULS

ONE'S PERCEPTION OF DEPTH, TRIGGERING A SORT OF VERTIGO. THE THIRD DIMENSION IS ALSO MUDDLED BY MIRAGES, WHICH OBLITERATE DISTANCES, MAKING THE SIGHT OF VERY DISTANT OBJECTS SUCH AS AN ISLAND, MOUNTAIN CHAIN, SHIP, THE SEA OR A VILLAGE SEEM CLOSE AT HAND.

THE DIMENSION OF TIME FADES AWAY IN POLAR SUMMERS AND WINTERS, WHEN EVEN AT MIDNIGHT THE SUN DOES NOT DESCEND BELOW THE HORIZON AND DAYS SEEM TO BE SUSPENDED IN PERPETUAL LIGHT.

THE FIFTH DIMENSION IS THAT OF DREAMS, SO TO SPEAK, WHICH CONTINUOUSLY PRODUCE THE MOST MAGICAL VISIONS THAT THE POLAR REGIONS HAVE IN STORE FOR AMAZED OBSERVERS: SOLAR HALOES,

The Ice Kingdom

Introduction

MOCK SUNS AND MOCK MOONS, THAT IS, THE MAGICAL APPEARANCE OF TWO, THREE AND SOMETIMES FOUR SUNS AND MOONS AT THE SAME TIME, THE TRICK OF A DIVINITY THAT SEEMS TO ENJOY POKING FUN AT HUMANS. THEN THERE ARE THE AURORAS, THE SOUTHERN AND NORTHERN LIGHTS, WHICH APPEAR SYMMETRICALLY IN THE SOUTHERN AND NORTHERN HEMISPHERE, CONSISTING OF A PALE "DRAPE" OF FLOWING LINES DELICATELY COLORED GREEN, RED AND PINK. THOSE FORTUNATE ENOUGH TO SEE THESE MAGIC DISPLAYS OF LIGHT MAY BE LED TO BELIEVE THAT THESE DRAPERIES TOUCH THE HORIZON AND COME FROM A DISTANT WORLD.

503 ● Weddell Sea (Antarctica) - Midnight sun.

504-505 ● Argentina - Perito Moreno Glacier.

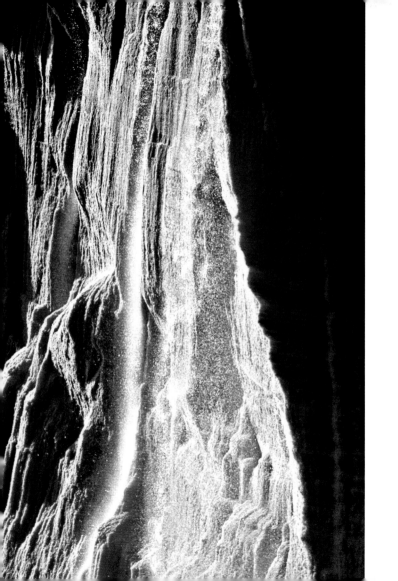

506-507 ● Svalbard Islands
(Norway) - Ice falls on the
Austfonna Glacier.

508-509 ● Churchill
(Canada) - Polar bears
tackle a deep crack
in the ice.

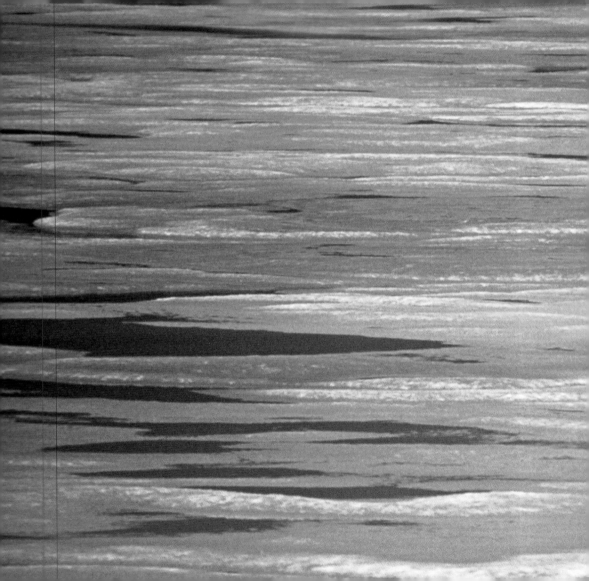

510-511 • Victoria Land (Antarctica) - Iceberg off the Adare Peninsula.

512-513 ● Adelia Land (Antarctica) - Dumont d'Urville Glacier.

514-515 ● Ross Sea (Antarctic) - Ross Ice Shelf.

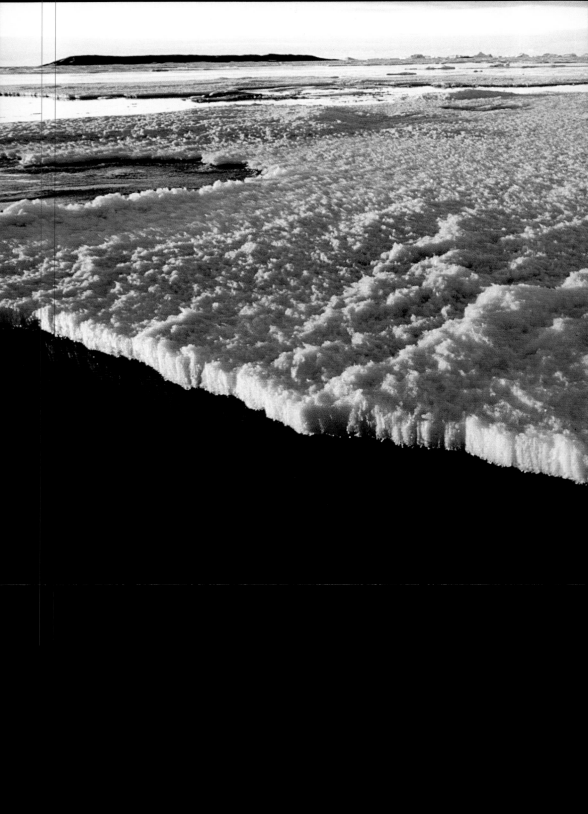

516-517 ● Antarctic Peninsula (Antarctica) - Iceberg off Cuverville Island.

518-519 ● Greenland - Midnight sun over Disko Bay.

520-521 ● Antarctic Peninsula (Antarctica) - Tower iceberg near Anvers Island.

522-523 ● Weddell Sea (Antarctica) - Iceberg.

524-525 ● Ross Sea (Antarctic) - Ice-pack in the Gerlache Strait.

521

526-527 ● Greenland - Disko Bay.

528-529 ● Falkland Islands (United Kingdom) - Iceberg off Saunders Island.

530-531 ● South Atlantic Ocean - Flock of petrels over an iceberg.

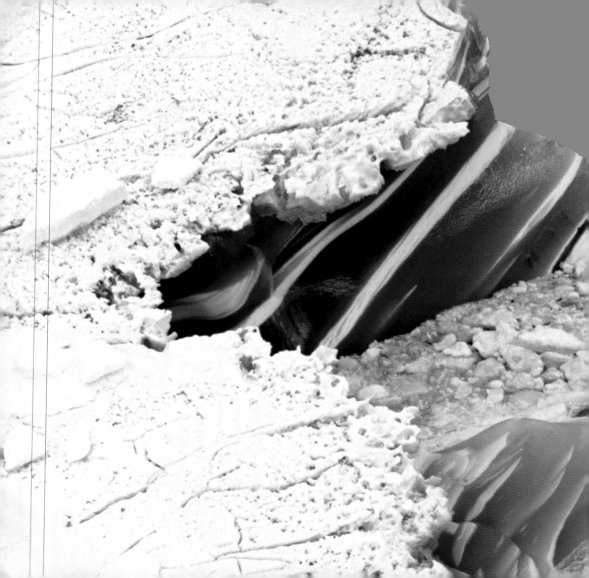

The WORLD of STONE

Utah-Arizona (USA) - Monument Valley National Park.

INTRODUCTION The World of Stone

THE SUN RISES OVER THE VAST MESAS OF THE AMERICAN WEST LIKE AN OMNIPOTENT DEITY. AFTER A NIGHT OF BITTER COLD, THE FIRST RAYS REPRESENT THE REVIVAL OF LIFE, BLOOD THAT FLOWS AGAIN PAST THE LONG SHADOWS CREATED BY THE GIGANTIC STONES. AS IF WE WERE AT THE DAWN OF HUMAN EXISTENCE, IMMENSE STONE ARCHES, TALL AND STEEP WALLS, BIZARRE SHAPES, SHARP CRAGS, AND THIN STRIPS OF SANDSTONE, LIMESTONE, SCHIST AND GRANITE DESCRIBE A FANTASTIC SCENE INTERRUPTED BY RARE SIGNS OF MODERN LIFE: A DIRT ROAD, A SOLITARY ROAD SIGN, THE DUST FROM A CAR PASSING IN THE DISTANCE, THE WISPY TRAIL LEFT BY A JET PLANE IN THE SKY. A FEW INSECTS NUMBED BY THE NIGHT COLD LOOK FOR A HID-

Sinai (Egypt) - Wadi Khudra.

INTRODUCTION

ING PLACE, ESCAPING THEIR FAMISHED PREDATORS. THE AIR IS CLEAR AND ALL THE COLORS SEEM TO BE KINDLED BY AN UNREAL LIGHT AND THE STREAKED ROCKS – BE THEY SMOOTH OR VARIEGATED, STRATIFIED OR SCULPT-ED, ERODED OR SLATEY – ARE FLAMING RED, PINK, OCHER AND YELLOW. THIS IS THE IRIDESCENT PALETTE THAT GENERATED THE EPICS OF THE NAVAHO AND HOPI, OF THE MYSTERIOUS ANASAZI CLIFF DWELLERS SUCH AS THE PUEBLO, AND OF THE PIONEERS AND COWBOYS. IN THE SOLITUDE OF THESE VAST EXPANSES, CONTEMPLAT-ING BARE CLIFFS BEARING CARVINGS TENS OF THOU-SANDS OF YEARS OLD, MAN CAN FIND HIMSELF OR, IF HE PREFERS, GOD. IT IS NO ACCIDENT THAT UPON VIEWING THE GRAND CANYON, THE NATURALIST DONALD CUL-

INTRODUCTION The World of Stone

ROSS PEATTIE STATED HE HAD FELT "THE LORD'S WILL." OTHER HAVE DESCRIBED THE GORGES CARVED BY THE COLORADO RIVER AS THE "LAST JUDGMENT OF NATURE." BECAUSE OF THE GRANDIOSITY OF ITS SCENERY, THE AMERICAN SOUTHWEST IS THE "STONE GARDEN" *PAR EXCELLENCE.* OTHER CANYONS, DESERTS AND ROCK FORMATIONS IN DIFFERENT CORNERS OF THE EARTH BOAST FANTASTIC PANORAMAS. IN AUSTRALIA, CERTAIN GEOLOGICAL FORMATIONS, INCLUDING AYERS ROCK, HAVE ASTOUNDED THE WORLD AND PROMPTED LEADING ARTISTS TO GRAPPLE WITH THE MYSTERY AND FASCINA-TION OF THESE PHENOMENA. THE TROPICAL REGIONS OF AFRICA OFFER THE CRAGGY NEEDLES OF THE SAHARA RANGE, THE EXTRAORDINARY TASSILI ROCK PAINTINGS,

The World of Stone

Introduction

THE GRANITE TOWERS OF HOGGAR AND THE FALAISE DOGON, AN IMPRESSIVE CLIFF THAT CUTS THROUGH THE TROPICAL STEPPE IN MALI: ROCKS WITH FANTASTIC SHAPES, WILD VALLEYS AND RARE, PRECIOUS SPRINGS. IN THE MEDITERRANEAN REGION THE GORGES OF THE RHONE OR THE FABULOUS "FAIRIES' CHIMNEYS" IN CAPPADOCIA LOOK LIKE MONUMENTS THAT NATURE BUILT FOR HERSELF, WITH THE SOLE AIM OF ASTONISHING HUMAN BEINGS. IN THESE PLACES, THE GEOLOGICAL ERAS AND THE HISTORY OF LIFE ITSELF ARE CARVED ON THE ROCKS LIKE PAGES OF A PRECIOUS ILLUMINATED MANUSCRIPT THAT BEARS CORALS, SHELLS AND FOSSILS MOUNTED LIKE JEWELS.

539 • New Mexico (USA) - Shiprock.

540-541 • Utah-Arizona (USA) - Monument Valley National Park.

542-545 ● Utah-Arizona (USA) - Different aspects of *buttes* and *mittens* in Monument Valley National Park.

546 ● Utah-Arizona (USA) - Monument Valley National Park, *mitten*.

547 ● Utah (USA) - Arches National Park, Delicate Arch.

548-549 ● Arizona (USA) - Grand Canyon, South Rim from Pima Point.

550-551 • Arizona (USA) - Grand Canyon, Yavapai Point area.

552-553 • Utah (USA) - Bryce Canyon, east edge of the Paunsaugunt Plateau.

554-555 • Utah (USA) - Bryce Canyon.

556-557 • Utah (USA) - Castle Rock, La Sal Mountains.

South Dakota (USA) - Badland National Park.

Arizona (USA) - Antelope Canyon, sandstone eroded by seasonal floods

Colorado (USA) - erosion caused by the Colorado River.

THE PLANET IS FORMED OF ROCK, CREATED BY THE FORCES OF NATURE: THE WIND, THE RAIN AND FIRE. THIS HAS CREATED AUTHENTIC WORKS OF ART, STONE SCULPTURES WHOSE FORMS AND PROPORTIONS HAVE ENCHANTED MAN FOR TENS OF THOUSANDS OF YEARS.

564-565 ● Region of Koilou (Congo) - 'Gorges de Diosso', Pointe Noire.

566-567 ● Tanzania - Rift Valley.

568-569 ● Region of Tamanrasset (Algeria) - Hoggar massif.

570-571 ● Namibia - Fish River Canyon.

Province of Segovia (Spain) - Hoz del Duratón.

574-575 ● Jordan -
Jebel Harun.

576-577 ● Sinai (Egypt) -
Forest of Columns,
Jebel Fuga.

578-579 ● Cappadocia
(Turkey) - Countryside
around Uchisar.

580-581 ● Southern Gobi
(Mongolia) - The mountains
of Khermiyn-Tsau.

582-583 ● Yunnan (China) - Clay Forest near Yuanmou.

584-585 ● Western Australia (Australia) - Joffre Falls in Karijini National Park.

586-587 ● Western Australia (Australia) -
Bungle-Bungle Range, Purnululu
National Park.

588-589 ● Northern Territory (Australia) -
Ayers Rock.

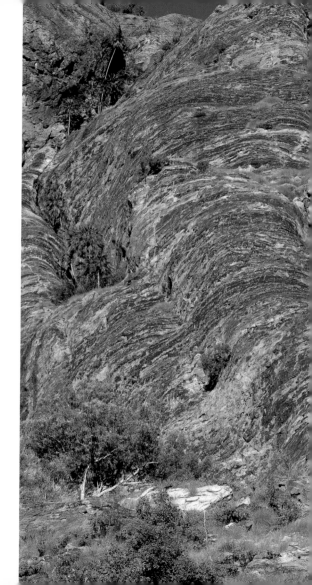

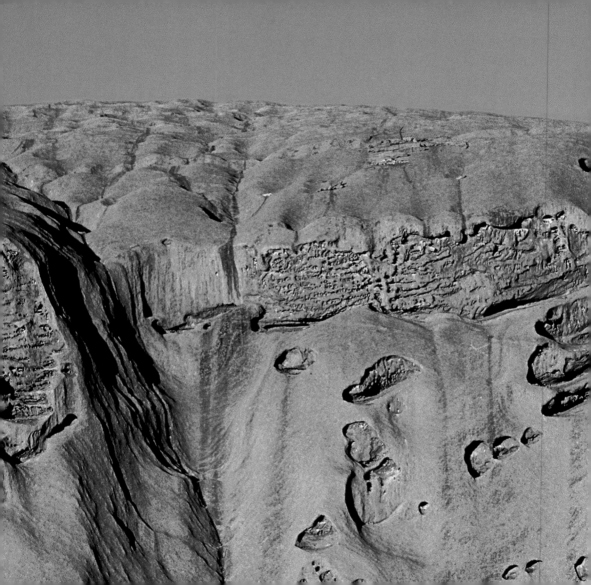

The BORDERS BETWEEN

Tanzania - Baobab in Tarangire National Park.

INTRODUCTION The World in Between

THERE IS A CORNER OF OUR PLANET WHERE TIME HAS STOOD STILL AND, IN ENVIRONMENTS MARKED BY WILD BEAUTY, SCENES DATING FROM PRIMEVAL ERAS ARE REPEATED AGAIN AND AGAIN: A LEOPARD USES HER MUZZLE TO MOVE HER CUB OUT OF DANGER; A PEACEFUL HERD OF ELEPHANTS AMBLES TOWARD THE FOOT OF THE HILLS; A GROUP OF LIONESSES AMBUSH AN UNWARY BUFFALO … THE CRADLE OF HUMANITY, ACCORDING TO PALEOLONTOLOGICAL FINDS, IS ALSO THE PRIMAL NOAH'S ARK, A PLACE IN WHICH LIFE IN ALL ITS FORMS FOUND THE IDEAL CONDITIONS FOR DEVELOPMENT. AT THE EDGE OF THE TROPICAL FORESTS AND EQUATORIAL JUNGLES, FAR FROM THE GREAT DESERTS, THE AFRICAN SAVANNAS ARE THE

INTRODUCTION

RICHEST ENVIRONMENT, AND NO OTHER REGION IN THE WORLD CAN BOAST SUCH A WEALTH OF FAUNA. CROSSED FROM NORTH TO SOUTH BY THE DEEP FISSURES OF THE RIFT VALLEY, EAST AFRICA IS A VERITABLE PATCHWORK OF LANDSCAPES, A SITE WHERE THE DAYS SEEM TO HAVE NO TIME AND THE NIGHTS ARE MARKED BY THE BLOOD-CURDLING HOWLS OF HYENAS AND ROARS OF LIONESSES. WHAT IS IT THAT THE SERENGETI NATIONAL PARK IN TANZANIA HAS IN COMMON WITH THE NATIONAL PARKS OF WEST AUSTRALIA, THE VENEZUELAN *LLANOS* OR THE BRAZILIAN *CAMPOS*? FIRST OF ALL, THE PRIMACY OF GRASSY PLANTS, THAT IS, THE VAST PRAIRIES IN WHICH THE GRASS IS SO HIGH THAT IT HIDES THE PREDATORS FROM THE VIEW OF THEIR PREY, AND

The World in Between
Introduction

THE ONLY PLANTS ABLE TO GROW AND SURVIVE IN THE TORRID HEAT ARE THORNY ACACIAS OR BAOBAB WITH THEIR TENTACULAR ROOTS. IN EVERY PART OF THE WORLD, SAVANNAS ARE THE LAND OF RUNNING ANIMALS. IN FACT, THEY ARE THE HOME OF SOME OF THE FASTEST CREATURES IN THE WORLD: CHEETAHS AND OSTRICHES IN AFRICA, NANDU IN SOUTH AMERICA, AND EMUS IN AUSTRALIA. A SWATHE OF THE LOST PARADISE, THE SAVANNAS HAVE INSPIRED ARTISTS AND WRITERS. BUT THEY ARE ABOVE ALL THE SYMBOLS OF NATURE AND THE MAIN CHARACTERS IN THE PLAY THAT IS PERFORMED EVERY DAY ON THIS STAGE, THE LIVING ICONS OF ECOLOGICAL CONSERVATION.

595 • Australia - Patterns created by vegetation in the semi-desertic Uluru National Park.

596-597 • Kenya - Column of migrating gnu in Masai Mara National Reserve.

598 ● Tanzania - Herds of zebra
and gnu in Ngorongoro
Conservation Area.

598-599 ● Tanzania - Aerial view
of Serengeti National Park.

" BURNED BY THE SUN, THE SAVANNA IS TURNED INTO AN IMMENSE GREEN CARPET AFTER THE STORMS, WHICH FLOOD THE LAND WITH RUSHING WATER, FILL THE WATERHOLES, SWELL THE DRIED RIVERBEDS, AND PERMEATE THE AIR WITH NEW SMELLS AND NEW LIGHT. "

600 ● Kenya - Elephants drinking in Tsavo East National Park.

601 ● Tanzania - Giraffes beneath an umbrella-shaped acacia in Serengeti National Park.

602-603 • Kenya - Cheetah lying in wait in the tall grass in Masai Mara National Reserve.

604-605 ● Tanzania - Grassland with shrubs in Serengeti National Park.

606-607 ● Tanzania - Acacia scrubland in Ngorongoro Conservation Area.

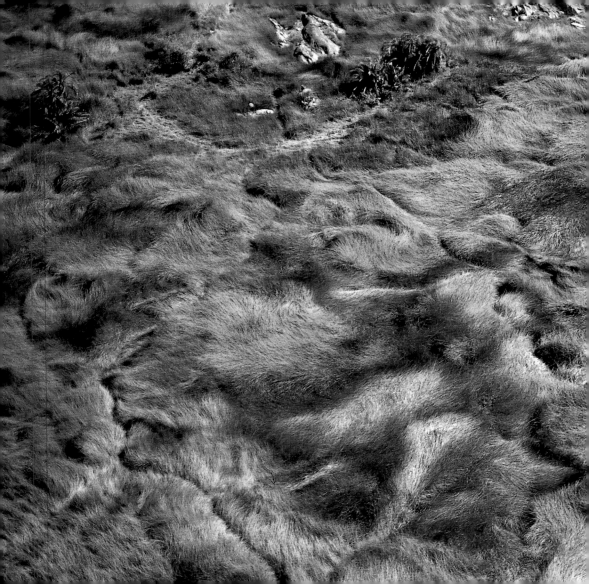

Argentina - Patagonian pampas and (in the background) the Andes, Los Glaciares National Park.

610-611 • Chile - Herd of guanacos, Torres del Paine National Park

612-613 ● Australia - Dunes with spiny bushes along the Canning Stock Route.

614-615 ● Australia - Ayers Rock, Ayers Rock-Mount Olga National Park.

INVISIBLE
BOUNDARIES

Piedmont (Italy) - Snowy landscape.

INTRODUCTION Invisible Boundaries

The ROAR OF THE PLANE DROWNS OUT THE OTHER NOISES A THOUSAND METERS BELOW IT, ON THE PLAIN THAT EXTENDS AS FAR AS THE EYE CAN SEE. ONLY THE HORIZON, WITH ITS SLIGHT CURVE, SEEMS ABLE TO ENCOMPASS IT. THE REASSURING SHAPES OF THE CULTIVATED FIELDS RELIEVE THE VIEWER'S SENSE OF VERTIGO. THE NETWORK OF INTERSECTING CANALS AND ROADS CREATES MYSTERIOUS STREAKS ON THE LAND. IN THE MIDDLE, LIKE PIECES OF A PUZZLE, THE INDIVIDUAL PLOTS OF LAND HAVE DIFFERENT COLORS: LAND ABOUT TO BE SOWN IS OCHER, THAT JUST FERTILIZED IS DARK BROWN, WHILE THE PLOTS LEFT FALLOW ARE GREEN AND YELLOW. IN THE DISTANCE, THE INSISTENT GLEAMS OF THE SUN ARE REFLECTED IN A WATERWAY. PERHAPS THIS

INTRODUCTION Invisible Boundaries

PLAIN IS THE RESULT OF THOUSANDS OF YEARS OF ERO-
SION EFFECTED BY THAT RIVER, WHICH HAS INCESSANTLY
CARRIED AND DEPOSITED TONS AND TONS OF SEDIMENT
DOWNSTREAM.THE FLAT STRETCHES OF LAND ARE THE
PLACES WHERE MAN FOUND THE BEST CONDITIONS TO
SETTLE AND TO CREATE CIVILIZATION. IF THE CLIMATE IS
FAVORABLE, THERE ARE MANY RIVERS AND STREAMS
THAT PROVIDE WATER AND FISH; PLAINS ARE THE HOME
OF MOST PREDATORS AND LIVESTOCK; COMMUNICA-
TIONS ARE EASIER HERE AND BOTH CULTURE AND GOODS
CAN TRAVEL RAPIDLY. WHILE MOUNTAINS AND THE SEA
EMBODY SPIRITUALITY AND BOLDNESS RESPECTIVELY,
PLAINS STAND FOR REASON, THE STRAIGHT LINE, GEOME-
TRY. YET ART HAS BEEN GREATLY INSPIRED BY THE HAR-

MONIC ALTERNATION OF COLORS AND FORMS AND OF CULTIVATED AND FALLOW FIELDS, IN VAST, OPEN SPACE. EXAMPLES OF SUCH ART INCLUDE THE MASTERPIECES OF RELIGIOUS AND CIVIC ARCHITECTURE FROM THE ROMAN AQUEDUCTS TO GOTHIC CATHEDRALS AND THE MOST AUDACIOUS SKYSCRAPERS, AND THE DREAMLIKE VISIONS OF VAN GOGH OR THE RURAL LANDSCAPES OF CASORATI. WHEN CLIMATIC CONDITIONS AND THE ALTITUDE ARE LESS BENIGN, LONG-TERM RESEARCH AND ADAPTATION ARE NEEDED IN ORDER FOR PLAINS TO SURVIVE. THESE ARE THE PLAINS WITH VARIOUS NAMES – THE SOUTH AMERICAN PAMPA, THE MEDITERRANEAN GARIGA, THE EURASIAN STEPPE, THE SOUTH AFRICAN VELD, AND THE HUNGARIAN PUSZTA – BARREN, ROCKY AREAS DOTTED

INTRODUCTION Invisible Boundaries

WITH SMALL BUSHES AND SHRUBS, IN WHICH THE RARE TRACKS ARE INTERRUPTED BY THE SLIMY WATER OF A FORD OR BY AN UNNEGOTIABLE STRETCH OF MARSHY LAND MUCH LIKE QUICKSAND THAT END IN VAST PRAIRIES WHERE ONE CAN TRAVEL FOR DAYS WITHOUT MEETING A SOUL. OR IT IS A MONOTONOUS TUNDRA COVERED WITH MOSS AND LICHEN, ITS SUBSOIL FROZEN ALL YEAR LONG AND ENDLESS SNOW-BLANKETED EXPANSES, PLACES WHERE THE AIR SEEMS TO BE RARIFIED AND THE WAIL OF THE WIND IS A FAITHFUL COMPANION FOR TRAVELERS. LANDS THAT ARE APPARENTLY HOSTILE BUT WITH THE MOST SUBTLE FASCINATION MAN CAN PERCEIVE: GOING AND BEING BEYOND THE CONFINES, BEYOND THE MOST REMOTE SHORES AND LAST SEA.

622-623 ● Ireland - Fields of poppies in County Kildare.

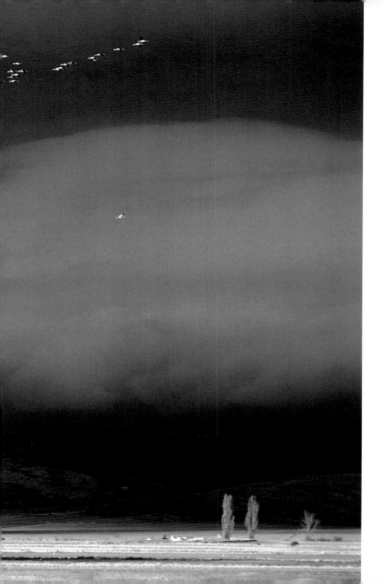

624-625 ● Saragossa (Spain) - Plain of Gallocanta.

626-627 ● Tanzania - Shrubs in the savannah, Serengeti National Park.

628-629 ● Tanzania - Aerial view of the northern regions, Serengeti National Park.

630-631 ● Alaska (USA) - Denali National Park.

632-633 ● California (USA) - Antelope Valley, California Poppy Reserve.

634-635 ● Wyoming (USA) - Pitchstone Plateau, Yellowstone National Park.

ROLLING HILLS

Washington, (USA) - Palouse Hills seen from Steptoe Butte State Park.

INTRODUCTION Rolling Hills

A PAINTING BY LEONARDO DA VINCI, WHICH IS UNJUSTLY FAMOUS ONLY BECAUSE OF THE AMBIGUOUS EXPRESSION OF THE WOMAN PORTRAYED, CONVEYS ALL THE FASCINATION OF TUSCANY, WHICH IS EPITOMIZED IN THE SINUOUS LINES AND TENUOUS COLORS THE MASTER USED TO RENDER THE BACKGROUND. LEONARDO MAY HAVE CHOSEN THAT LANDSCAPE FOR "TECHNICAL" REASONS, SINCE ITS OUTLINES ENHANCE THE SOFT FORMS OF THE MONA LISA AND SOMEHOW ALSO EMBODY HER SUBTLY EROTIC OVERTONES. IN OTHER WORDS, THE TUSCAN HILLS HERE BECOME A SYMBOL OF FEMININITY. OR THE ARTIST MAY HAVE SIMPLY RECALLED THE HILLOCKS, THE ROWS OF CYPRESSES, THE SOLITARY OAKS AND THE OLIVE TREES HE HAD

ALWAYS SEEN SINCE HE WAS A BOY AND RODE THROUGH THE COUNTRYSIDE ON A MULE. AGAIN, THERE MAY HAVE BEEN NO OTHER POSSIBLE SETTING FOR THE WORLD'S MOST FAMOUS PAINTING: WAVES OF REDDISH EARTH ALTERNATING WITH OTHER WAVES OF GREENERY AND, LIKE BUOYS LOST IN THE SEA, TREES, FARMSTEADS AND TOWNS WITH TURRETED WALLS. A LANDSCAPE THAT IN SPRING TAKES ON EVERY COLOR, THE PINK OF PEACHES, THE WHITE OF DAISIES, THE RED OF POPPIES, THE YELLOW OF SUNFLOWERS ....

THE HILLS OF NEW ENGLAND ARE QUITE DIFFERENT, WITH THEIR THICK, EVER-CHANGING WOODS AND THE INDIAN SUMMER THAT TRANSFORMS THEM INTO FIERY FORESTS. THE HILLS OF NEW ZEALAND ARE SO DISTANT

Rolling Hills
Introduction

YET SO NEAR, A MOONSCAPE, LIKE THE ENGLISH HILLS ON THE OTHER SIDE OF THE GLOBE, BUT WITH A GREEN SO RICH IT SEEMS TO HAVE BEEN PAINTED. THE BARREN HILLS BELOW THE ANDES THAT BEGIN IN PATAGONIA SEEM WILD, SILHOUETTED AGAINST THE LOW LIGHT THAT PENETRATES THE CRYSTAL-CLEAR AIR AT THAT LATITUDE. AND THE GREEN HILLS OF AFRICA SEEMED SO EXOTIC TO WESTERN WRITERS THAT THEY FELL IN LOVE WITH THE BLACK CONTINENT.... SO DIFFERENT AND YET SO ALIKE, IN EVERY PART OF THE WORLD, IN FORM AND COLOR, FROM THOSE THAT WERE PART OF OUR CHILDHOOD OR THAT WE WERE TAUGHT TO LOVE.

641 ● Tibet (China) - The hills near Shigatse.

642-643 ● Tuscany (Italy) - Val d'Orcia.

644-645 • Umbria (Italy) - The Sybilline Mountains.

646-647 • Tuscany (Italy) - The crags of Volterra.

648-649 • Marches (Italy) - Apennine foothills in Umbria and the Marches.

650-651 • Alsatia (France) - The upper Rhine.

652-653 • Tanzania - Ruaha National Park.

654 ● South Australia (Australia) - North Flinders Ranges.

655 ● Northern Territory (Australia) - Undulating hills in MacDonnel Ranges National Park.

656-657 ● Arizona (USA) - Saguaro National Park.

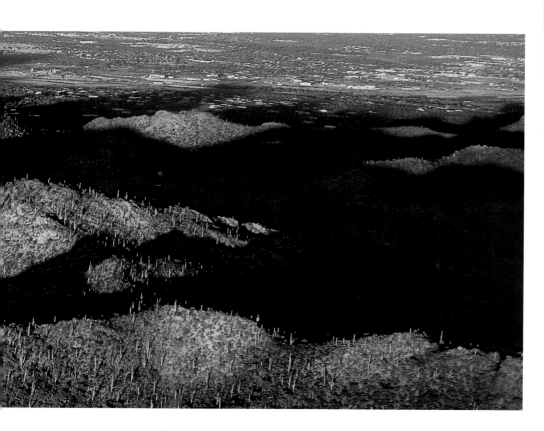

658-659 ● Utah (USA) - Bryce Canyon National Park.

Between

LAND and WATER

- Louisiana (USA) - Bayou forest.

INTRODUCTION Between Land and Water

THE FIRST MAN WHO VENTURED INTO A MARSH MUST HAVE THOUGHT HE HAD ENTERED HELL: A GLOOMY UNDERWORLD INHABITED BY FIERCE PREDATORS, INFESTED BY FOUL SMELLS AND BY THE CONTINUOUS, UNBEARABLE HUM OF INSECTS; AN ENVIRONMENT SUFFOCATED BY OVERWHELMING VEGETATION, MONSTERS, QUICKSANDS AND MALARIA. IN FACT, MARSHES HAVE ALWAYS TRIGGERED NIGHTMARISH FANTASIES IN THE SOUL OF THOSE WHO HAVE HAD TO TRAVERSE THEM, SUCH AS THE FIRST EXPLORERS OF THE NILE. IN THE HEART OF AFRICA, THESE HARDY MEN WERE SEEKING THE FABULOUS MOUNTAINS OF THE MOON, WHERE THEY BELIEVED THE WELLSPRINGS OF THE DIVINE WATER ROSE, BUT THEIR SEARCH CAME TO AN ABRUPT HALT WHEN THEY ARRIVED

INTRODUCTION Between Land and Water

AT IMPASSABLE SWAMPS BEYOND LAY ONLY MYSTERY. AND YET, SWAMPS HAVE ALSO BEEN GENEROUS TO MAN, PROVIDING WATER, EDIBLE FRUITS, WOOD, CANE AND RUSHES, FISH AND GAME BIRDS. THE RELATIONSHIP BETWEEN MAN AND MARSHLAND IS MARKED BY THE FOLLOWING CONTRADICTION: THE UNHEALTHIEST ENVIRONMENT IS ALSO THE WEALTHIEST IN OFFERING THE FRUIT OF VARIED FORMS OF LIFE TO THOSE WHO KNOW HOW TO EXPLOIT THEM. AND THIS IS NOT A QUESTION ONLY OF MATERIAL GOODS RESULTING FROM RECLAMATION, AS THESE MARSHY AREAS ARE ALSO A TREASURE TROVE OF MARVELOUS SCENERY AND A BIOLOGICAL PARADISE FOR THOUSANDS OF LIVING SPECIES. IN THOSE ZONES WHERE RECLAMATION HAS FAILED OR HAS NOT

Between Land and Water
Introduction

BEEN ATTEMPTED, THE MARSHLAND HAS PRESERVED THAT AURA OF MYSTERY AND FASCINATION REPRESENTED IN MANY WORKS OF LITERATURE, ART AND FILM: FOG PASSING OVER THE STAGNANT WATER OF THE SCOTTISH PEAT-BOGS; WILD HORSES GALLOPING ALONG THE CAMARGUE SWAMPS; THE FLAMINGOES ALIGHTING REGALLY AFTER A SHORT FLIGHT AMONG THE TROPICAL REEDS; THE SNAPPING OF THE ALLIGATORS, LORDS OF THE EVERGLADES; THE CRY OF THE BLUE HERON THAT RESOUNDS IN THE BOGS OF CANADA. IN THESE PLACES THE MARRIAGE OF LAND AND WATER HAS GENERATED A FABULOUS REALM, AS OPAQUE AS AN UNCUT DIAMOND AND AS PRECIOUS AS A PEARL IN ITS SHELL....

665 • New Caledonia (French Overseas Territories) - A 'heart' created by a mangrove swamp.

666-667 • South Carolina (USA) - Sea Islands.

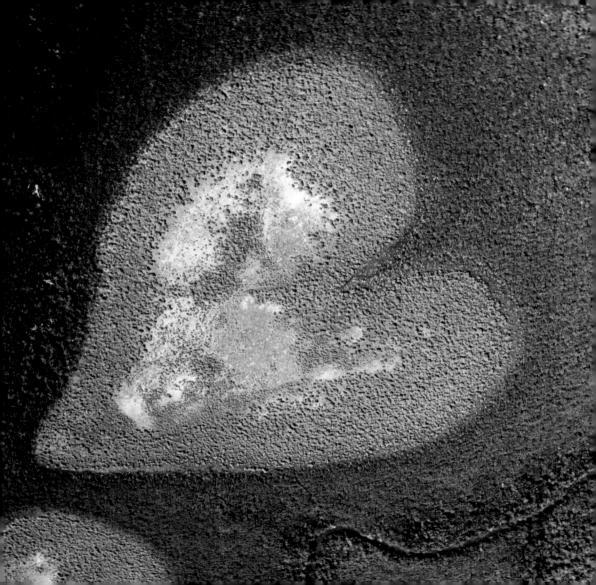

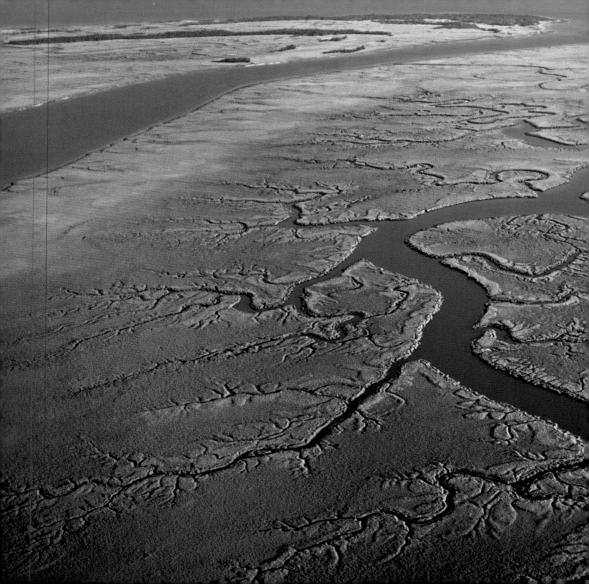

668-669 • Florida (USA) - Everglades National Park.

670-671 • Louisiana (USA) - Devil's Swamps, New Orleans.

672-673 • Northwest Territories (Canada) - Northwest shore of the Great Slave Lake.

674-675 • Northwest Territories (Canada) - Thaw in the tundra at Cape Bathurst.

676 ● Northern Territory (Australia) - Swamps in Magela Creek, Kakadu National Park.

677 ● Northern Territory (Australia) - Malaleuca forest, Kakadu National Park.

678-679 ● Northern Territory (Australia) - Kakadu National Park.

680-681 ● Northern Territory (Australia) - Aerial view of Djarr Djarr Wetlands, Kakadu National Park.

Botswana - Zebras in the marshes of the Okavango.

● Botswana - Hippopotamus in the marshes of the Okavango.

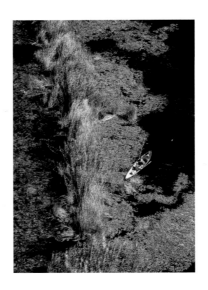

686-689 ● Egypt - Marshland around Lake Manzala, in the Nile Delta.

690-691 ● Ireland - Pools in County Mayo.

" WHITE HORSES GALLOP THROUGH THE SPRAY. GRAY HERONS AND GIANT PELICANS GLIDE BETWEEN THE RUSHES. AN ALLIGATOR STRETCHES OUT AMONG THE AQUATIC PLANTS. THE MARSHES ARE WHERE LAND AND WATER MEET, THEY ARE THE WET UNDERBELLY OF THE PLANET AND TEEM WITH LIFE. "

692-693 ● Provence (France) - Horses in the Camargue.

694-695 ● Provence (France) - Marshes along the Rhone.

The RAINBOW PLANET

Utah (USA) - Waterfalls in Lower Calf Creek.

INTRODUCTION The Rainbow Planet

MILLIONS OF YEARS AGO, MAGICAL VIBRATIONS BEGAN TO COLOR THE TIMELESS DAYS OF SPACE. NOW THE COSMOS IS A SHIMMERING KALEIDOSCOPE OF STARS, PLANETS AND GASSES WITH MYRIAD NUANCES. FROM ITS WOMB THE EARTH WAS BORN, A MAGICAL SPHERE IN WHICH ONE CAN READ THE PAST, PRESENT AND FUTURE OF THE UNIVERSE, THE CRADLE OF LIFE AS WE KNOW IT AND THE PLACE WHERE BEAUTY WAS FIRST CONSCIOUSLY CONTEMPLATED. THE WELLSPRINGS OF LIFE HAVE IRRIGATED THE INERT MINERALS OF A PLANET LOST AMID BILLIONS OF OTHER PLANETS AND STARS, TRANSFORMING IT INTO AN IRIDESCENT GLOBE, MUCH LIKE A CHILD'S TOY. UNIQUE AND EXTRAORDINARY, THIS EARTH IS A THREE-DIMENSIONAL CANVAS TRAVERSED BY

INTRODUCTION The Rainbow Planet

THE BRUSHSTROKES OF A GREAT ARTIST. THE THOU-
SANDS OF NUANCES OF WHITE ON THE GLACIERS THAT
ENFOLD THE POLAR REGIONS, CROWN THE HIGHEST
PEAKS, AND SLIDE SILENTLY INTO THE OCEANS; THE
WHITE OF THE CLOUDS AND SEAGULLS THAT STAND OUT
AGAINST THE SKY; THE WHITE OF CORAL SAND AND OF
THE BLINDING REFLECTIONS OF THE SUN. THE INFINITELY
VARIED HUES OF BLUE: THE SKY-BLUE OF THE AIR AT ANY
LATITUDE, THE TURQUOISE IN TROPICAL SEAS, THE SAP-
PHIRE OF ALPINE LAKES, THE COBALT BLUE IN DEEP WA-
TER .... THEN THERE ARE THE REDS AND ORANGES OF
BLISTERING SAND, OF MAPLE LEAVES IN AUTUMN, OF THE
PLUMAGE OF CERTAIN EQUATORIAL BIRDS, OF SUNSETS
ILLUMINATED BY THE LAST RAYS OF THE SUN, OF THE

The Rainbow Planet

Introduction

MAGMA THAT FLOWS MENACINGLY FROM A VOLCANO. THE EVANESCENT GREEN OF THE AFRICAN FORESTS IMMERSED IN VELVETY CLOUDS, THE BRIGHT GREEN OF IRISH MEADOWS AND IN CULTIVATED FIELDS IN MAY, THE RICH GREEN OF LIZARDS IN THE SUNLIGHT AND DARK GREEN OF PINE GROVES IN WINTER. NEXT COME THE YELLOWS, FROM THE OCHER OF THE HILLS AROUND SIENA TO THE RUSTY HUES OF FERRUGINOUS MOUNTAINS, THE GOLD IN RIPE WHEAT AND THE BLOND SHADES IN THE COATS OF LARGE FELINES. LASTLY, THERE IS THE DARK BLACK IN THE OCEAN ABYSSES, IN MOONLESS NIGHTS, IN THE INFINITE SPACE THAT ENFOLDS OUR PLANET.

701 • Iceland - Kirkjubaejarklaustur Delta.

702-703 • Ethiopia - Lichens, Bale Mountains National Park.

704-705 • Alberta (Canada) - Salt flats in Wood Buffalo National Park.

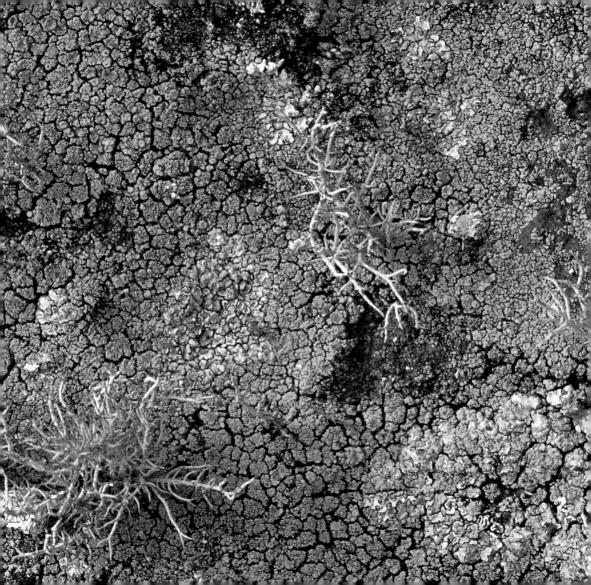

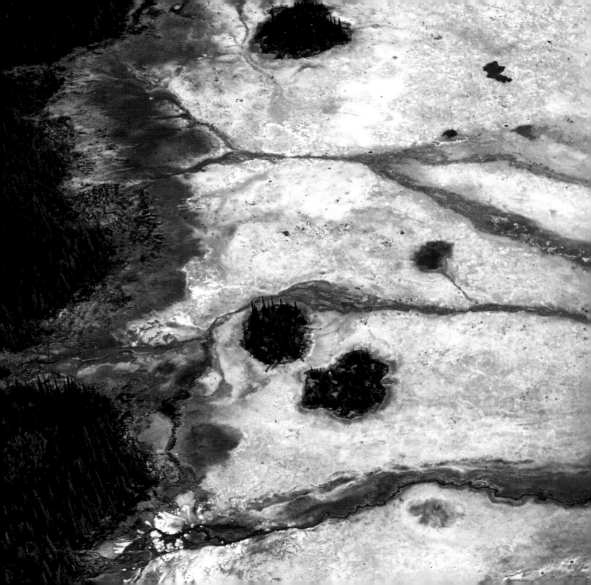

Baja California (Mexico) - Dunes on the beach of Scammons Lagoon.

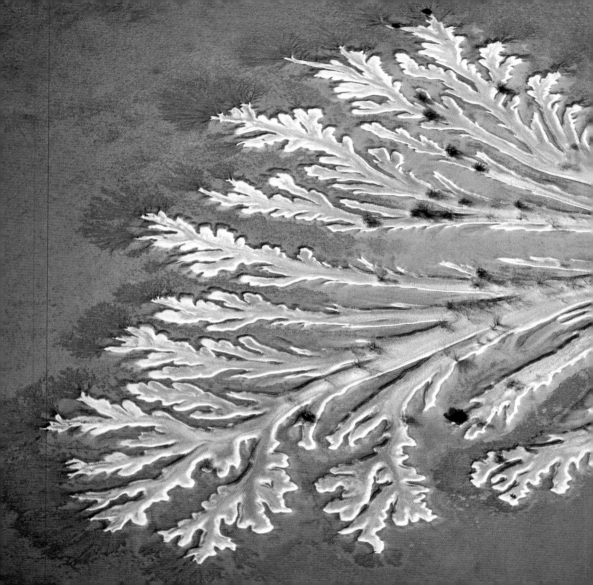

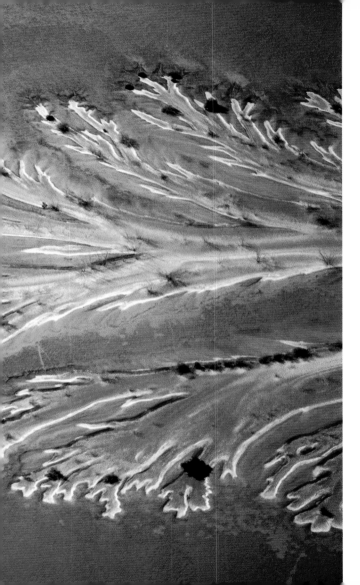

● Sonora (Mexico) - Erosion produced in the desert sand by ancient water flows.

710-711 ● Umbria (Italy) - Fields at Castelluccio, near Norcia.

712-713 ● Lazio (Italy) - Fields of poppies, sunflowers and alfalfa on the Sybilline Mountains.

714-715 ● England (United Kingdom) - Rows of poplars and fruit trees.

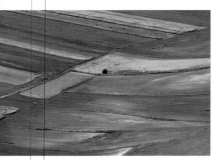 

" FIELDS LIKE FORMS IN CONTINUAL MOVEMENT, LIKE PATCHES OF COLOR ON A PALETTE: HERE THE WHITE OF THE DAISIES, THERE THE RED OF THE POPPIES, IN BETWEEN THE VIOLET OF THE LAVENDER AND THE YELLOW OF THE SUNFLOWERS. ALL AROUND, GREEN IN ALL ITS HUES SPREADS LIKE A WATERCOLOR OR IS CLUMPED IN THICKETS LIKE A MODERN PAINTING. "

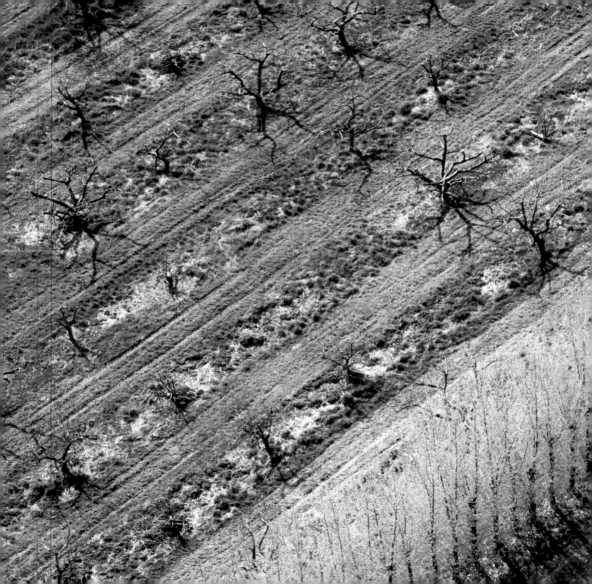

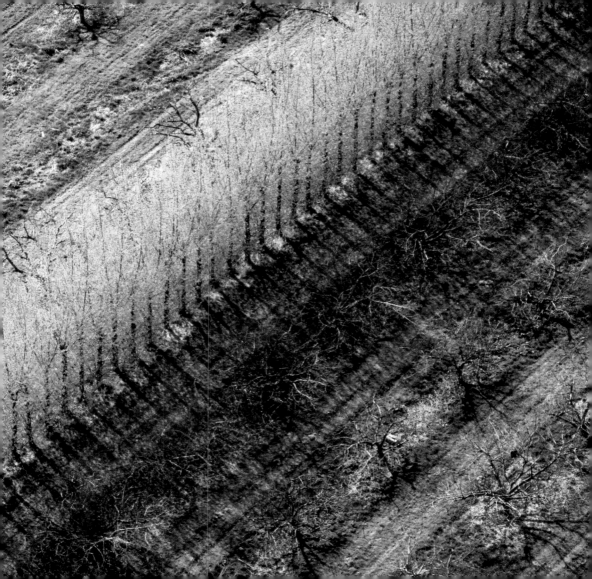

716-717 ● Landeyjar (Iceland) - Iron deposits along a river Delta.

718-719 ● Tuamotu Islands (French Polynesia) - Rangiroa.

720-721 ● Iceland - Water plays among the springs.

722-723 ● Oregon (USA) - Basalt columns in the ravine formed by the River Umpqua.

724-725 ● Algeria - light, colors and forms in the Sahara Desert.

726-727 ● California (USA) - Salt formations in San Francisco Bay.

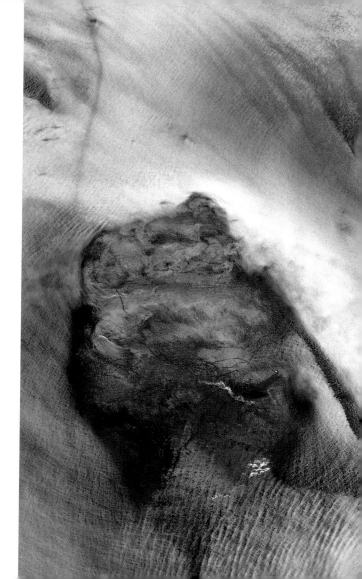

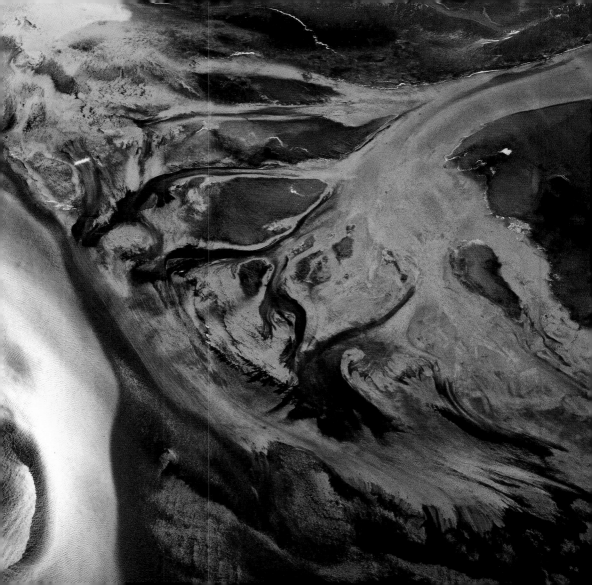

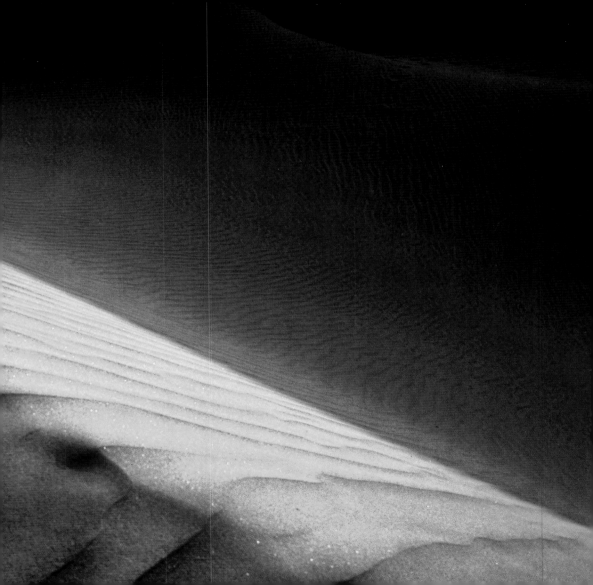

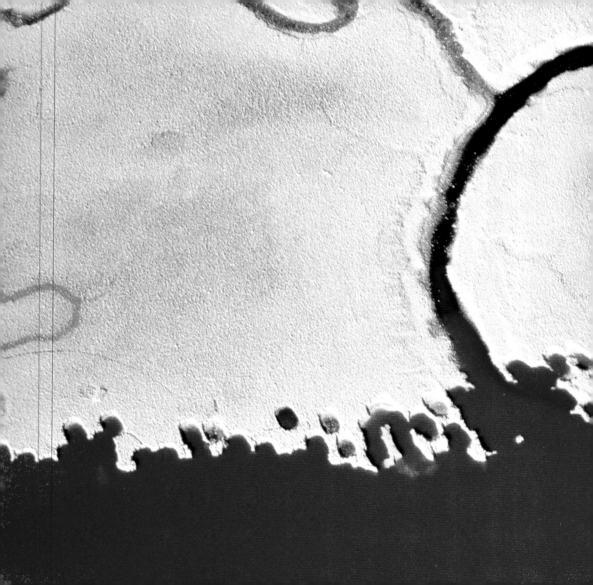

INDEX

AUTHOR
Biographies

ALBERTO BERTOLAZZI

Born in 1961, Alberto Bertolazzi studied Philosophy at the University of Pavia. He is an enthusiastic traveler and naturalist who, following a brief period as a teacher, began writing for various Italian newspapers, including *La Repubblica*, *La Stampa*, *Il Giornale Nuovo*, *Il Piccolo di Trieste* and *Il Giorno*, and the periodicals *Meridiani*, *Sestante* and *Panorama*. For White Star Publications he has written the volumes *Lisbon* (1997) and *Portugal* (1998) and has contributed to the editing of many other books dedicated to the natural world.

AFGHANISTAN
Band i Amir Lakes, 444-445.

ALGERIA
Hoggar Massif, 568-569.

ANTARCTICA
Adare Peninsula, 510-511.
Anvers Island, 520-521.
Cuverville Island, 516-517.
Fief Mountains, 122-123.
Gerlache Strait, 524-525.
Dumont D'Urville Glacier, 512-513.
Ross Ice Shelf, 514-515.
Weddel Sea, 28-29, 522-523, 503.
Wiencke Island, 122-123.

ARGENTINA
Cerro Torre, 116-117.
Fitz Roy, 114-115.
Parque Nac. Los Glaciares, 114-115, 608-609
Patagonian pampas, 608-609.
Perito Moreno, 504-505.

AUSTRALIA
Ayers Rock, 588-589, 614-615.
Boult Reef, 158.
Bungle-Bungle Range, 586-587.
Canning Stock Route, 612-613.
Djarr Djarr Kreek, 680-681.
East Gippsland Errinundra National Park, 348, 348-349.
Everard Ranges, 285.
F. L. Gordon Wild Rivers National Park, 351.
Great Ocean Road, 154-155.
Joffre Falls, 584-585.
Kakadu National Park, 677, 678-679.
Lady Musgrave Island, 159.
Lake Boomanjin, 442.
Lake McKenzie, 442-443.
MacDonnel Ranges, 655.
Magela Kreek, 676.
Mt. Field Nat. Park, 350.
North Flinders Ranges, 654.
Pelorus River Scenic Res., 352-353.
Simpson Desert, 286-287.
Strzelecki Desert, 284.
The Pinnacles, 288-289.
Uluru National Park, 595.
Wall of China, 231.
Whitsunday Island, 156-157.

BAHAMAS
Exuma Keys, 214-215.

BOTSWANA
Okavango, 682-685.

BRAZIL
Anavilhanas Archipelago, 400-401.
Iguazú Falls, 26-27, 402-403.
Rio Negro, 400-401.

CANADA
Cape Bathurst, 674-675.
Great Slave Lake, Yellowknife, 430, 430-431, 672-673.
Kluane Lake, 421, 428-429.
Laurentian Highlands, 422-423.
Moraine Lake, 434.
Mount Kidd, 424-425.
Niagara Falls, 390-391, 391.
O'Hara Lake, 435.
Peace River, 392.
Peyto Lake, 436-437.
Polar bears, 508-509.
Rampart Range Peaks, 98-99.
Rocky Mountains, 100-101.
Slave River, 392-393.
St. Elias Mountains, 88-89.
Wedge Pond, 424-425.
Wood Buffalo National Park, 704-705.

CHILE

Lake Pehoe 440-441.
Torres del Paine, 112-113, 610-611.

CHINA

Brahmaputra, 412-413, 414-415.
Everest, 66-67.
Huanglong, 30-31.
Kailash, 75.
Lake Shuzheng, 446-447.
Lake Yamdruk, 446-447.
Shigatse, 641.
Siguniang, 64-65.
Silk Road, 282-283.
Yuanmou Clay Forest, 582-583.

CONGO

Congo River, 386-387.
Gorges de Diosso, Pointe Noire, 316-317, 564-565.
Kouilou River, 310-311, 382-383.
Loufoulakari Falls, 384-385.
Mayombe Forest, 6-7, 312-313.

COSTARICA

Foresta di Monteverde, 334-335.
Tortuguero National Park, 336-337.

CUBA

Cayo Coco, 210-211.

DENMARK

Disko Bay (Greenland), 518-519, 526-527.
Nordskoven Forest, 306-307.

EGYPT

Ain Um Ahkmed Oasis, 227.
Alexandria, 178-179.
Bahariya, 254/257.
Eastern Sahara, 266-267.
Faiyum Oasis, 266-267.
Farafra, 256-257, 260-261.
Forest of Columns, Jebel Fuga, 576-577.
Gobal Straits, 166-167.
Jebel Bayda, 452-453.
Jebel Musa, 268-269.
Lake Manzala, 686-689.
Lake Nasser, 450-451.
Lake Siwa, 452-453.
Nile, 388-389, 686-689.
Ras Gharib, 1.
Shali, 264, 264-265.
Siwa Oasis, 262-263, 264, 264-265.
Tiran Island, 168-169.
Wadi Khudra, 535.
White Desert, 258/261.

ETHIOPIA

Bale Mountains National Park, 702-703.

FINLAND

Oulanka Joki River, 368-369.
Oulanka National Park, 368-369.

FRANCE

Aiguille Noire, 52-53.
Camargue, 692-693.
Corsica, 196-197.
Falaise d'Aval, 200-201.
Finistère, 198-199.
Gironde, 370, 370-371.
Golfe d'Etretat, 200-201.
Mont Blanc, 50/55.
Pays de Caux, 200.
Petit Langoustier, 197.
Porquerolles, 197.
Rhine River, 694-695.
Upper Rhine, 650-651.

FRENCH OVERSEAS TERRITORIES

Blue Lagoon 2-3.
Bora Bora, 129, 140-141, 142-143.
Huahiné, 150-153, 344-345.
Marquesas Islands, 138-139.
Mataiva, 148-149.
New Caledonia, 665.
Nuku Hiva, 138-139.
Piton de la Fournaise, 494-497.
Pointe du Chateau, 212-213.
Rangiroa 2-3, 144-145, 146-147, 346-347, 718-719.
Society Islands, 129, 140-141, 142-143, 150-153, 344-345, 346-347.
Tuamotu, 144/149, 718-719.

GERMANY

Bavaria, 60-61, 304-305.
Elbe River, 364-365.
Little Watzmann, 60-61.
Watzmann, 60-61.

GREECE

Crete, 180-181.
Elaphonisos, 180-181.

INDIA

Bagirathi Peaks, 82-83.
Shivling, 82.

INDONESIA

Gunung Leuser National Park, 339.

IRELAND

County Kildare, 622-623.
County Mayo, 690-691.
Gweebarra River, 372/375.
Sligo Bay, 206/209.

ICELAND

Grimsvotn, 482-483.
Helgafell, 482.
Kirkjubaejarklaustur, 701.
Landeyjar, 716-717, 720-721.
Vestmannaeyjar, 482.

ISRAEL

Negev, 270-271.

ITALY

Balze di Volterra, 646-647.

INDEX

Brenta, 37, 42, 43.
Budelli, 133.
Cala delle Zagare, 187.
Castelluccio di Norcia, 710-711.
Catinaccio, 38-39.
Chiaia di Luna, 188-189.
Cima Madonna, 40-41.
Cima Tosa, 42.
Crozzon di Brenta, 43.
Dente del Gigante, 48-49.
Dolomites, 37/43.
Brenta Dolomites, 4-5.
Eruption of Etna, 471, 484-485.
Faraglione di Pizzomunno, 186.
Fusine River, 362-363.
Forest of Tarvisio, 300-301.
Island of Capraia, 190-191.
Island of Vulcano, 189.
Lampedusa, 182-183.
Maddalena, 184-185.
Maremma, 302-303.
Monte Bianco, 48-49.
Monte Cervino, 46-47.
Monte Rosa, 44-45.
Laghi di Fusine National Park, 464-465.
Piedmont, 616.
Pink Beach, 133.
Ponza, 188-189.
Rabbit Beach, 182-183.
Sass Maor, 40-41.
Sibilline Mountains, 644-646, 712-713.
Val d'Orcia, 642-643.
Valsesia, 296-297.

JORDAN
Jebel Harun, 574-575.
Wadi Rum, 272-273.

KENYA
Lake Magadi, 462-463.
Lake Nakuru, 16-17.
Masai Mara Nat. Reserve, 596-597, 602-603.
Tsavo East National Park, 600.

LIBYA
Akakus zoomorphic rocks, 240-241.
Fezzan, 25.
Libyan Desert, 232-239.
Sebha Oasis, 242-243.

MALAYSIA
Kinabalu National Park, 342-343.
Sabah Forest, 340-341.

MEXICO
Las Coloradas, 218-219.
Manzanillo, 216-217.
Scammons Lagoon, 706-707.
Sonora, 708-709.

MICRONESIA
Caroline Islands, 136-137.
Lekes Sandspit, 134-135.
Palau, 136-137.

MONGOLIA
Khermiyn-Tsau, 580-581.

Southern Gobi Desert, 580-581.

NAMIBIA
Bakers Bay, 176-177.
Fish River Canyon, 570-571.
Gamsberg Pass, 248-249.
Namibian Desert, 244/247, 252-253.
Skeleton Coast, 174-175.
Sussusvlei, 250-251.

NEPAL
Ama Dablam, 19.
Annapurna, 76-77.
Everest, 68-69, 70-71.
Gangapurna, 74.
Lhotse, 72-73.
Machapuchare, 78-79.
Nuptse, 70-71.
Tent Peak, 80-81.

NEW ZEALAND
Bowen Falls, 408-409.
Coromandel State Forest Park, 354-355.
Mount Cook, 126-127.
Southern Alps, 124-125.
Sutherlands Falls, 408.

NIGERIA
Niger River, 291, 308-309, 379.

NORWAY
Alattiojoki River, 366-367.

Austfonna Glacier, 506-507.
Kautokeino, 366-367.

PAKISTAN
Gasherbrun IV, 79.
K2, 32, 86-87.
Masherbrun, 84-85.

PAPUA NEW GUINEA
Karawari River, 410-411.

PERU
Cordillera Blanca, 118/121.
Tigre River, 404-405.
Monte Huandoy, 120.

PHILIPPINES
Palawan, 162-163.

PORTUGAL
Praia da Rocha, 195.
Sagres (Algarve), 194-195.

RUANDA
Kagera River, 376-377, 377.

SEYCHELLES
La Digue, 160-161.

SPAIN
Gallocanta Plateau, 624-625.
Hoz del Duratón, 572-573.
Isola di Conejera, 192-193.
Monte Perdido, 62-63.

Ordesa National Park, 62-63.
Pyrenees, 62-63.

SRI LANKA
Udowattakele Forest Reserve, 338-339.

SWEDEN
Lake Marsliden, 468-469.
Lapland, 298-299.

SWITZERLAND
Eiger, 56-57, 58-59.
Jungfrau, 56-57.
Mönch, 56-57, 58-59.

TANZANIA
Bagamoyo, 170-171.
Dar Es Salaam, 172-173.
Kilimanjaro, 478-479.
Lake Natron, 14-15, 458-459, 460-461.
Lake Victoria, 318-319, 454-455.
Ngorongoro Conservation Area, 598, 606-607.
Oldonyo Lengai, 480-481.
Rift Valley, 566-567.
Ruaha National Park, 652-653.
Rubondo Island, 318-319.
Rufiji River, 378-379.
Serengeti National Park, 456-457, 598-599, 601, 604-605, 626-627, 628-629.

Tarangire National Park, 591.

THAILAND
Rai Lai Beach, 164-165.

TURKEY
Uchisar, 578-579.

UNITED KINGDOM
Bishop Rock, 204-205.
Gower, 202-203.
Lake District of Cumbria, 466-467.
Saunders Island, 528-529.
Tarne Hows Lakes, 466-467.
West Glamorgan, 202-203.

USA
Alaska Range, 94-95.
Antelope Canyon, 560-561, 561.
Antelope Valley, 632-633.
Aspen, 320-321, 324-325.
Badwater Pool, 276-277.
Badlands National Park, 558-559.
Banded Peak, 322-323.
Bayou Forest, 660.
Big Sur, 220-221.
Bryce Canyon, 554-555, 658-659.
Castle Rock, 556-557.

Colorado River, 280-281, 396-397, 562-563.
Copper River Valley, 13.
Delicate Arch, 547.
Denali National Park, 630-631.
Devil's Swamps, 670-671.
Everglades National Park, 668-669.
Glacier Bay National Park and Preserve, 92-93.
Grand Teton National Park, 102-103.
Grand Prismatic Spring, 736.
Half Dome, 108-109.
Kalaupapa Nat. Historical Park, 398-399.
Kilauea, 475, 486-487, 488-489, 492-493.
Lake Jackson, 432, 432-433.
Lake Powell, 417, 438-439.
Lake Woodruff Nat. Wildlife Refuge, 330-333.
Lower Calf Creek, 697.
Mauna Ulu, 493.
Monument Valley, 8-9, 533, 540-541, 542-545, 546.
Moosehead Lake, 426-427.
Mt. Rainier, 105.
Mt. St. Helens, 476-477.
Mt. McKinley, 90-91.
Napali Coast, 224-225.
Olympic Nat. Park, 104.
Palouse Hills, 636.
Paunsaugunt Plateau, 552-553.

Pitchstone Plateau, 634-635.
Redwood Natural Park, 326-327, 328-329.
Saguaro National Park, 656-657.
San Francisco Bay, 726-727.
Sea Islands, 666-667.
Sequoia and Kings Canyon National Park, 110-111.
Shiprock, 539.
Sonora Desert, 278-279.
South Rim, 548-549.
Superstition Mountains, 278.
Umpqua River, 722-723.
Waialeale, Kauai, 399.
Waipio Bay 222-223.
Wrangell-St. Elias National Park, 96-97.
Yavapai Point, 396-397, 550-551.
Yellowstone National Park, 11, 20.
Yellowstone River, 394-395, 395.
Yosemite National Park, 106-107.
Zabriskie Point, 274-275.

VENEZUELA
Angel Falls, 357.
Orinoco River, 406-407.

ZIMBABWE
Victoria Falls, 380-381.

PHOTO CREDITS

Pages 194-195, 195 Giulio Veggi/Archivio White Star

Pages 196-197, 197 Guido Alberto Rossi/ The Image Bank

Pages 198-199 Livio Bourbon/Archivio White Star

Pages 200, 200-201 Guido Alberto Rossi/ The Image Bank

Pages 202-203, 204-205 Getty Images/ Laura Ronchi

Pages 206-207, 208, 208-209, 210-211 Antonio Attini/Archivio White Star

Pages 212-213, 213 Marcello Bertinetti/Archivio White Star

Pages 214-215 Yann Arthus-Bertrand/ Corbis/Contrasto

Pages 216-217 David Alan Harvey/ Magnum Photos/Contrasto

Pages 218-219 George D. Lepp/Corbis/ Contrasto

Pages 220-221, 222-223, 223, 224-225 Antonio Attini/Archivio White Star

Page 227 Marcello Bertinetti/Archivio White Star

Page 231 Dave G. Houser/Corbis/ Contrasto

Pages 232-233, 234-235, 236-237, 238-239, 240-241, 241, 242-243 Gianni and Tiziana Baldizzone/Archivio White Star

Pages 244-245 Alamy

Pages 246, 246-247 Berndt Fischer

Pages 248-249 Yann Arthus-Bertrand/ Corbis/Contrasto

Pages 250-251 Jan Tove

Pages 252-253 Yann Arthus-Bertrand/ Corbis/Contrasto

Pages 254, 255, 256-257, 258, 259, 260-261, 262-263, 264, 264-265, 266, 267 Marcello Bertinetti/Archivio White Star

Pages 268-269 Antonio Attini/Archivio White Star

Pages 270-271 Itamar Grinberg/Archivio White Star

Pages 272-273 Massimo Borchi/Archivio White Star

Pages 274-275, 276-277 Marcello Bertinetti/Archivio White Star

Page 278 Tim Fitzharris

Pages 278-279 Massimo Borchi/Archivio White Star

Pages 280-281 Tim Fitzharris

Pages 282-283 Liu Liqun/Corbis/ Contrasto

Pages 284, 285, 286-287 Jean-Paul Ferrero/Auscape

Pages 288-289 Michael Jensen/Auscape

Page 291 Marcello Bertinetti/Archivio White Star

Page 294 Yann Arthus-Bertrand/Corbis/ Contrasto

Pages 296-297, 298-299 Giulio Veggi/ Archivio White Star

Pages 300-301 Luciano Ramires/Archivio White Star

Pages 302-303 Marcello Bertinetti/ Archivio White Star

Pages 304-305, 305 Sven Zellner

Pages 306-307 Jan Tove

Pages 308-309, 310-311, 312, 312-313, 314-315, 316-317, 318-319 Marcello Bertinetti/Archivio White Star

Pages 320-321 Darrell Gulin/Corbis/ Contrasto

Pages 322-323, 324-325 Jim Wark

Pages 326, 327 Tim Fitzharrris

Pages 328-329 Gary Braasch/Corbis/ Contrasto

Pages 330-331 Getty Images/Laura Ronchi

Pages 332-333 Jan Tove

Pages 334-335 Gary Braasch/Corbis/ Contrasto

Pages 336-337 André Bartschi

Pages 338-339, 339 Elio Della Ferrera

Pages 340-341 Martin Harvey/NHPA

Pages 342-343 Gunter Ziesler

Pages 344-345, 346-347 Marcello Bertinetti/Archivio White Star

Page 348 Jaime Plaza Van Roon/ Auscape

Pages 348-349 D. Parer & E. Parer-Cook/ Auscape

Page 350 Jaime Plaza Van Roon/Auscape

Page 351 Dennis Harding/Auscape

Pages 352-353 Jan Tove

Page 353 Tom Till/Auscape

Pages 354-355 Michael s. Yamashita/ Corbis/Contrasto

Page 357 James Marshall/Corbis/Contrasto

Pages 362-363 Luciano Ramires/Archivio White Star

Pages 364-365 Wolfgang Kaehler/ Corbis/Contrasto

Pages 366-367 Farrell Grehan/Corbis/ Contrasto

Pages 368-369 Hautala Hannu/Panda Photo

Pages 370, 370-371 Guido Alberto Rossi/The Image Bank

Pages 372, 373, 374-375, 375 Antonio Attini/Archivio White Star

Pages 376-377, 377, 378-379, 379 Marcello Bertinetti/Archivio White Star

Pages 380-381 Torleif Svensso/Corbis/ Contrasto

Pages 382-383, 384, 384-385, 386-387, 388-389 Marcello Bertinetti/Archivio White Star

Pages 390-391 Ron Watts/Corbis/ Contrasto

Page 391 Wolfgang Kaehler/Corbis/ Contrasto

PHOTO CREDITS

Pages 558-559 Antonio Attini/Archivio White Star

Pages 560-561 Dave G. Houser/Corbis/Contrasto

Page 561 Darrell Gulin/Corbis/Contrasto

Pages 562-563 Owaki-Kulla/Corbis/Contrasto

Pages 564, 565, 566-567 Marcello Bertinetti/Archivio White Star

Pages 568-569 Tiziana and Gianni Baldizzone/Corbis/Contrasto

Pages 570-571 Nigel J. Dennis; Gallo Images/Corbis/Contrasto

Pages 572-573 Juan Carlos Munoz/Agefotostock/Contrasto

Pages 574-575 Massimo Borchi/Archivio White Star

Pages 576-577 Antonio Attini/Archivio White Star

Pages 578-579 Massimo Borchi/Archivio White Star

Pages 580-581 Konstantin Mikhailov

Pages 582-583 Liu Weixiong/Panorama Stock

Pages 584-585 S. Wilby & C. Ciantar/Auscape

Pages 586, 586-587 Jean-Paul Ferrero/Auscape

Pages 588-589 Paul A. Souders/Corbis/Contrasto

Page 591 Francois Savigny/Nature Picture Library

Page 595 Yann Arthus-Bertrand/Corbis/Contrasto

Pages 596-597 Fritz Polking; Frank Lane Picture Library/Corbis/Contrasto

Pages 598, 598-599 Marcello Bertinetti/Archivio White Star

Page 600 Alamy

Page 601 Marcello Bertinetti/Archivio White Star

Pages 602-603 Paul A. Souders/Corbis/Contrasto

Pages 604-605, 606-607 Marcello Bertinetti/Archivio White Star

Pages 608-609 Craig Lovell/Corbis/Contrasto

Pages 610-611 Theo Allofs/Corbis/Contrasto

Pages 612-613 Jean-Paul Ferrero/Auscape

Pages 614-615 Mark Laricchia/Corbis/Contrasto

Page 617 Marcello Bertinetti/Archivio White Star

Pages 622-623 Richard Cummins/Corbis/Contrasto

Pages 624-625 Vincent Munier

Pages 626-627, 628-629, 629 Marcello Bertinetti/Archivio White Star

Pages 630-631 Ron Sanford/Corbis/Contrasto

Pages 632-633 Liz Hymans/Corbis/Contrasto

Pages 634-635 Jason Hawkes

Page 637 Darrell Gulin/Corbis/Contrasto

Page 640 Panorama Stock

Pages 642-643, 644, 644-645, 646-647 Giulio Veggi/Archivio White Star

Pages 648-649 Andrea Rontini

Pages 650-651 Vincent Munier

Pages 652-653 Marcello Bertinetti/Archivio White Star

Pages 654, 655 Jean-Paul Ferrero/Auscape

Pages 656-657, 658-659 Antonio Attini/Archivio White Star

Page 661 Olivier Cirendini/Lonely Planet Images

Page 665 Mesner Patrick/Gamma Presse/Contrasto

Pages 666-667 Jason Hawkes

Pages 668-669, 670-671 Antonio Attini/Archivio White Star

Pages 672-673 Jim Wark

Pages 674-675 Hans Strand

Pages 676, 677 Jean-Paul Ferrero/Auscape

Pages 678-679 S. Wilby & C. Ciantar/Auscape

Pages 680-681 Jean-Paul Ferrero/Auscape

Pages 682-683 Alamy Images

Pages 684-685 Alamy

Pages 686, 687, 688-689 Marcello Bertinetti/Archivio White Star

Pages 690-691 Antonio Attini/Archivio White Star

Page 692 Tom Brakefield/Corbis/Contrasto

Page 693 Jason Hawkes

Pages 694-695 Royalty-Free/Corbis/Contrasto

Page 697 Darrel Gulin/Corbis/Contrasto

Page 701 Hans Strand

Pages 702-703 Elio Della Ferrera

Pages 704-705, 706-707, 708-709 Jim Wark

Pages 710 a sinistra, 710 a destra, 711 Andrea Rontini

Pages 712-713 Renato Fano and Annamaria Flagiello

Pages 714-715 Jason Hawkes

Pages 716-717 Hans Strand

Pages 718-719 Marcello Bertinetti/Archivio White Star

Pages 720-721 Hans Strand

Pages 722-723 Gary Braasch/Corbis/Contrasto

Pages 724-725 Tiziana and Gianni Baldizzone/Archivio White Star

Pages 726-727 Getty Images/Laura Ronchi

Page 736 Antonio Attini/Archivio White Star

Cover Marcello Bertinetti/Archivio White Star

Back cover Alfio Garozzo/Archivio White Star

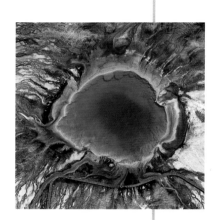

Wyoming (USA) - Grand Prismatic Spring, Yellowstone National Park.